Event-based state-feedback control of physically interconnected systems

Dissertation zur Erlangung des Grades
eines Doktor-Ingenieurs
der Fakultät für Elektrotechnik und Informationstechnik
an der Ruhr-Universität Bochum

von
Christian Stöcker
geboren in Salzkotten

Bochum, 2014

Gutachter: Prof. Dr.-Ing. J. Lunze
Prof.dr.ir. W.P.M.H. Heemels

Dissertation eingereicht am: 14. Januar 2014
Tag der mündlichen Prüfung: 12. März 2014

Bibliografische Information der Deutschen Nationalbibliothek

Die Deutsche Nationalbibliothek verzeichnet diese Publikation in der Deutschen Nationalbibliografie; detaillierte bibliografische Daten sind im Internet über http://dnb.d-nb.de abrufbar.

ISBN 978-3-8325-3682-4

Logos Verlag Berlin GmbH
Comeniushof, Gubener Str. 47,
10243 Berlin
Tel.: +49 (0)30 42 85 10 90
Fax: +49 (0)30 42 85 10 92
INTERNET: http://www.logos-verlag.de

Acknowledgements

This thesis is the result of five years of research at the Institute of Automation and Computer Control of Prof. Jan Lunze at the Ruhr-University Bochum. This period has been a challenging yet highly interesting and instructive time for me.

I would like to express my deepest gratitude to Prof. Jan Lunze for his confidence and his excellent advice and support all over the time. His well-founded knowledge on system theory and exceptional ability to make complex subjects easily understandable helped me to broaden my view on my own work.

I also thank Prof. Maurice Heemels for accepting to review this thesis. He is one of the leading experts in the field of event-based control and it's an honor to acknowledge the influence his outstanding publications had on my research.

Moreover, I would like to thank my partners in the DFG research project "Analysis and design of event-based control with quantized measurements - networked systems -" Manuela Sigurani, Prof. Lars Grüne, Prof. Oliver Junge and Dr. Péter Koltai. In several project meetings these persons showed me new interesting mathematical ways of looking at control theory.

Special thanks go to my colleges at the Institute of Automation and Computer Control: Andrej Mosebach, Axel Schild, Prof. Christian Schmid, Daniel Lehmann, Daniel Vey, Fabian Just, Melanie Schmidt, Michael Ungermann, Ozan Demir, René Schuh, Sebastian Drüppel, Sven Bodenburg, Thorsten Schlage and Yannick Nke. Thank you for stimulating discussions, proofreading my thesis and a relaxed and fun working atmosphere. I'm more than happy to say that some of these persons became good friends within the last years.

I would like to extend my gratitude to my parents Angelika and Wilfried and my brother Tim for their support over so many years.

Finally, I would like to thank the most important person in my life, my wife Joanna. Thank you for your patience, encouragement and enduring support. You are invaluable to me.

Bochum, March 2014 Christian Stöcker

Contents

Abstract

Event-based control is a means to restrict the feedback in control loops to event time instants that are determined by a well-defined triggering mechanism. The aim of this control strategy is to adapt the communication over the feedback link to the system behavior. Since events are generally triggered asynchronously in time, the event-based control strategy contrasts with the conventional control concepts where the feedback is continuous (continuous-time control) or triggered by an external clock (discrete-time control). In this thesis, a state-feedback approach to event-based control is extended from system with centralized sensor and actuator units to systems that are composed of physically interconnected subsystems.

The present work focuses on disturbance in interconnected systems, which is supposed to be best accomplished by a continuous state feedback. This consideration leads to the idea that the event-based state-feedback system shall be made to approximate the disturbance behavior of a continuous state-feedback system with adjustable precision. A dynamic model of the continuous control system is incorporated in the event-based controller in order to generate the control input and to determine the event times. By this means events are only triggered in response to a perturbation of the system.

The event-based state-feedback controller is considered as a networked controller that consists of local control units which can exchange information over a communication network. Three different methods for the event-based control of physically interconnected systems are investigated. In particular, decentralized, distributed and centralized state feedback is studied. It is shown that, depending upon the structure of the implemented state-feedback gain, the necessary communication effort varies from local point-to-point connections (unicasting) via the information transmissions from one sender to many receivers (multicasting) through to the distribution of information over the entire network (broadcasting).

The main results concern the design and analysis of event-based state-feedback control methods for physically interconnected systems which differ with respect to the model information in the control units and the communication effort. The stronger the interconnection between the subsystems is, the more model information is required in the control units and the higher is the necessary communication effort. For all approaches the disturbance behavior of a continuous state-feedback system is shown to be approximated with arbitrary accuracy by the event-based state-feedback system.

The novel event-based control methods are tested and evaluated through simulations and experiments on a thermofluid process implemented on a large-scale pilot plant. These investigations validate the theoretical results and, moreover, prove the proposed concepts to be robust with respect to model uncertainties. It is shown that event-based control adapts the feedback to the system behavior and, thus, considerably reduces the feedback communication effort compared to a conventional discrete-time control.

German extended abstract (Kurzfassung)

Ereignisbasierte Regelung

Die ereignisbasierte Regelung ist eine Methode, die zum Ziel hat, den Informationsaustausch zwischen den Komponenten eines Regelkreises über den Rückführzweig an das Verhalten der Regelstrecke anzupassen. Dabei werden nur dann Daten übertragen, wenn ein Ereignis die Notwendigkeit einer Informationsrückführung signalisiert, um ein gewünschtes Verhalten des Regelkreises sicherzustellen. Da die Ereignisauslösung im Allgemeinen in nicht-äquidistanten Zeitabständen erfolgt, unterscheidet sich das Konzept der ereignisbasierten Regelung wesentlich von einer herkömmlichen zeitdiskreten Regelung, bei welcher die Messung der Ausgangssignale der Regelstrecke und die Aktualisierung der Stellgrößen mit einer konstanten Abtastzeit passiert. Die bestehenden Methoden für den Regelungsentwurf und die Analyse des Regelkreisverhaltens lassen sich folglich nicht auf ereignisbasierte Regelungen übertragen und es sind daher neue Entwurfs- und Analysemethoden zu entwickeln.

Das Haupteinsatzgebiet ereignisbasierter Regelungsmethoden sind digital vernetzte dynamische Systeme. Dies sind Regelkreise, bei denen die Rückführung über ein digitales Kommunikationsnetzwerk geschlossen wird. Durch den Einsatz von ereignisbasierten Regelungen in diesen Systemen soll die Netzwerkauslastung auf ein für die Lösung der Regelungsaufgabe notwendiges Maß beschränkt und der Kommunikationsaufwand gegenüber einer zeitdiskreten Abtastregelung deutlich reduziert werden.

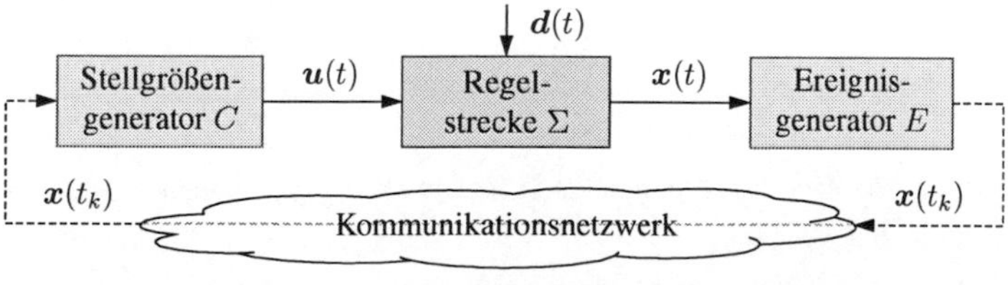

Abbildung 1: Ereignisbasierte Zustandsrückführung

Abbildung 1 zeigt die Struktur der ereignisbasierten Zustandsrückführung, auf der die in dieser Arbeit behandelten Methoden aufbauen. Der Regelkreis besteht aus der Regelstrecke Σ mit dem Stellsignal $\boldsymbol{u}(t)$, der Störung $\boldsymbol{d}(t)$ und dem Zustand $\boldsymbol{x}(t)$ und der ereignisbasierten Regelung, welche die folgenden Komponenten umfasst:

- Der *Ereignisgenerator* E bestimmt in Abhängigkeit vom Systemverhalten die Ereigniszeitpunkte t_k $(k = 0, 1, \ldots)$, zu denen der aktuelle Systemzustand $\boldsymbol{x}(t_k)$ an den Stellgrößengenerator gesendet wird.
- Der *Stellgrößengenerator* C erzeugt unter Verwendung der zu den Ereigniszeitpunkten t_k empfangenen Informationen $\boldsymbol{x}(t_k)$ die Stellgröße $\boldsymbol{u}(t)$.
- Das *Kommunikationsnetzwerk* wird genutzt, um den Datentransport vom Ereignisgenerator zum Stellgrößengenerator durchzuführen.

Die durchgezogenen Linien stellen eine kontinuierliche Informationsübertragung dar, während die gestrichelte Linie anzeigt, dass die Informationsrückkopplung nur zu den Ereigniszeitpunkten erfolgt.

Zielstellung der Arbeit

Eine wesentliche Voraussetzung für die Anwendung der ereignisbasierten Zustandsrückführung nach Abb. 1 ist, dass die Regelstrecke über eine zentrale Sensor- und Aktoreinheit verfügt, welche die vollständige Zustandsinformation $\boldsymbol{x}(t)$ ausgibt, bzw. über die alle Aktoren zentral angesteuert werden können. Viele technische Systeme erfüllen diese Voraussetzung jedoch nicht, da sie aus mehreren physikalisch gekoppelten Teilsystemen bestehen, die jeweils eigene Sensor- und Aktoreinheiten haben. Beispiele für solche Systeme sind große verfahrenstechnische Anlagen, Energieversorgungsnetze oder Kraft- und Luftfahrzeuge.

In dieser Arbeit wird der Ansatz zur ereignisbasierten Zustandsrückführung konzeptionell erweitert auf Systeme, die aus physikalisch gekoppelten Teilsystemen bestehen. Dabei liegt der Fokus auf der Untersuchung neuer Strukturen ereignisbasierter Regelungen, die in einer vereinheitlichten Form in Abb. 2 dargestellt sind. Die Regelstrecke Σ ist zusammengesetzt aus N Teilsystemen Σ_i $(i = 1, \ldots, N)$ mit der Stellgröße $\boldsymbol{u}_i(t)$, dem Zustand $\boldsymbol{x}_i(t)$ und der Störung $\boldsymbol{d}_i(t)$. Die Teilsysteme Σ_i sind über die Signale $\boldsymbol{s}_i(t)$ und $\boldsymbol{z}_i(t)$ miteinander verkoppelt. Der ereignisbasierte Regler F besteht aus den lokalen Reglereinheiten F_i, zwischen denen ein Informationsaustausch über das Kommunikationsnetzwerk erfolgt. Die untersuchten Strukturen der ereignisbasierten Regelung unterscheiden sich in dem Aufbau und in der Funktionsweise der Reglereinheiten F_i, sowie in der Topologie der Kommunikation zwischen diesen Komponenten. Gemein ist den Reglereinheiten F_i, dass sie jeweils einen Stellgrößengenerator und

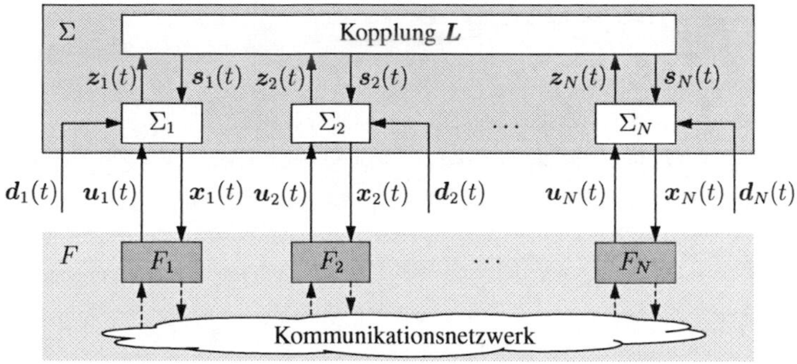

Abbildung 2: Vereinheitlichte Struktur der ereignisbasierten Regelung physikalisch gekoppelter Systeme

einen Ereignisgenerator beinhalten, die unter Verwendung nur lokal verfügbarer Informationen die Stellgröße $\boldsymbol{u}_i(t)$ erzeugen, bzw. die Ereigniszeitpunkte bestimmen.

Die Untersuchungen ereignisbasierter Regelungen basieren auf der grundsätzlichen Fragestellung, wann die Rückkopplung in einem Regelkreis geschlossen werden muss, um ein gegebenes Regelungsziel zu erfüllen. Für den Entwurf und die Analyse ereignisbasierter Regelungen für physikalisch gekoppelte Systeme lassen sich daraus die folgenden wichtigen Fragen ableiten:

- Wie sind die Komponenten des ereignisbasierten Reglers zu entwerfen?
- Wann müssen Informationen übertragen werden?
- Welche Informationen müssen zu den Ereigniszeitpunkten übertragen werden?
- Zwischen welchen Komponenten des ereignisbasierten Reglers ist ein Informationsaustausch notwendig?
- Wie sollen die empfangenen Informationen in den Komponenten des ereignisbasierten Reglers genutzt werden?

Die Beantwortung dieser Fragen hängt wesentlich vom Regelungsziel ab, was in dieser Arbeit die Störkompensation mit möglichst geringer Auslastung des Kommunikationsnetzwerks ist. Die behandelten Ansätze basieren auf der Grundidee, dass die ereignisbasierte Regelung das Störverhalten einer kontinuierlichen Zustandsrückführung (im Folgenden auch als Referenzsystem bezeichnet) mit einstellbarer Genauigkeit nachahmen soll.

Hauptergebnisse der Arbeit

In dieser Arbeit werden drei verschiedene Methoden zur ereignisbasierten Regelung physikalisch gekoppelter Systeme vorgeschlagen, die sich im Wesentlichen in zwei Aspekten voneinander unterscheiden: Erstens die in den Reglereinheiten verwendeten Modellinformationen und zweitens der Kommunikationsaufwand für den Datenaustausch zwischen den Reglereinheiten. Je stärker die Kopplungen zwischen den Teilsystemen sind, desto mehr Modellinformationen über das Gesamtsystem werden in den einzelnen Reglereinheiten zur Ermittlung der Stellgrößen und zur Bestimmung der Ereigniszeitpunkte benötigt und desto größer ist der Kommunikationsaufwand für den Informationsaustausch zwischen den Reglereinheiten.

Zunächst wird eine Bedingung an die Kommunikationstopologie für den Datenaustausch zwischen den Reglereinheiten in Abhängigkeit von der Struktur der im Referenzsystem verwendeten Rückführverstärkung angegeben, unter der die ereignisbasierte Regelung das Verhalten einer kontinuierlichen Zustandsrückführung mit einstellbarer Genauigkeit approximiert (Theorem 4.1). Ausgehend von diesem Ergebnis werden die drei Methoden zur ereignisbasierten Regelung physikalisch gekoppelter Systeme entwickelt.

Kapitel 5 behandelt eine Methode zur ereignisbasierten Regelung, die das Verhalten einer kontinuierlichen zentralen Zustandsrückführung nachahmt. Unter einer zentralen Regelung wird in diesem Zusammenhang verstanden, dass die Stellgröße $\boldsymbol{u}_i(t)$ für jedes Teilsystem eine Funktion des gesamten Systemzustands $\boldsymbol{x}(t)$ ist. Die Komponenten des ereignisbasierten Reglers beinhalten jeweils ein Modell der kontinuierlichen Zustandsrückführung. Der Informationsaustausch zwischen den Komponenten erfolgt zu den Ereigniszeitpunkten per Broadcast, so dass die ein Ereignis auslösende Reglereinheit F_i die Daten an alle übrigen Komponenten des ereignisbasierten Reglers übermittelt. Der Ansatz stellt eine direkte Erweiterung der zentralen ereignisbasierten Zustandsrückführung auf physikalisch gekoppelte Systeme dar, wobei die Teilsysteme beliebig stark miteinander verkoppelt sein können (Abschnitt 5.2.7).

Basierend auf dem zuvor beschriebenen Ansatz wird eine Methode zur ereignisbasierten Regelung physikalisch gekoppelter Systeme mit nicht vollständig messbarem Systemzustand vorgestellt. Es werden Systeme betrachtet, die so zerlegbar sind, dass der Zustand in einigen Teilsystemen verfügbar ist und in den übrigen Teilsytemen keine Informationen gemessen werden können. Ein neuer Algorithmus für den Entwurf der Komponenten des ereignisbasierten Reglers wird präsentiert, durch den garantiert wird, dass die ereignisbasierte Zustandsrückführung das Verhalten einer kontinuierlichen Regelung mit gewünschter Genauigkeit, trotz beschränkter Zustandsmessungen, approximiert (Abschnitt 5.4, Algorithmus 5.1).

Kapitel 6 untersucht eine Methode zur dezentralen ereignisbasierten Regelung, wobei die Reglereinheiten F_i dezentral und nur unter Verwendung der Modellinformationen des zugehörigen Teilsystems entworfen werden. Zu den Ereigniszeitpunkten erfolgt eine Informationsrückkopplung ausschließlich zwischen den Komponenten der Reglereinheit F_i, die das Ereignis ausgelöst hat, wohingegen keine Daten zwischen den Reglereinheiten ausgetauscht werden. Eine Methode zur Stabilitätsanalyse mit Vergleichssystemen führt auf eine Forderung nach schwacher Kopplung zwischen den Teilsystemen als Voraussetzung für die Anwendbarkeit dieses Ansatzes (Theorem 6.2).

Eine Methode zur verteilten ereignisbasierten Regelung wird in Kapitel 7 behandelt. Hierbei verwenden die Reglereinheiten Modellinformationen des zugehörigen Teilsystems und dessen Nachbarsysteme. Es wird ein neuer Mechanismus für die Ereignisgenerierung eingeführt, der neben dem Senden von Daten auch das Anfordern von Informationen von anderen Reglereinheiten initiiert (Abschnitt 7.3.4). Dadurch entsteht eine neue Art von ereignisbasierten Regelungen, die in der bestehenden Literatur bisher nicht betrachtet wurden. Der Ansatz ist geeignet für die Regelung von physikalisch gekoppleten Systemen, bei denen die Teilsysteme mit deren Nachbarsystemen stark verkoppelt sind.

Für alle untersuchten Strukturen werden die folgenden Eigenschaften nachgewiesen:

- Die ereignisbasierte Zustandsrückführung approximiert das Verhalten einer kontinuierlichen Zustandsrückführung mit einstellbarer Genauigkeit (Theoreme 5.1, 6.1, 7.3).
- Der Zustand der Regelstrecke wird durch die ereignisbasierte Zustandsrückführung in einer Umgebung des Ursprungs gehalten, deren Größe von den Entwurfsparametern der Regelungen abhängen (Theorem 6.4, Korollar 7.1).
- Die minimale Zeit, die zwischen zwei aufeinanderfolgenden Ereignissen vergeht, ist durch eine untere Schranke begrenzt (Theoreme 5.2, 6.6, 7.4, 7.5).

Weitere Beiträge dieser Arbeit bestehen in der Entwicklung neuer Methoden für

- die Stabilitätsanalyse gekoppelter ereignisbasierter Regelkreise (Theorem 6.2 für die dezentrale ereignisbasierte Regelung und Theoreme 7.1, 7.3 für die verteilte ereignisbasierte Regelung) unter Verwendung des *comparison principle* [113],
- die Schätzung von Kopplungseingangssignalen zur Reduzierung der Ereignisauslösung in der dezentralen ereignisbasierten Regelung (Abschnitt 6.2.5) und
- die ereignisbasierte Regelung physikalisch gekoppelter Systeme, wobei der Zustand einiger Teilsysteme nicht messbar ist (Abschnitt 5.4).

Alle vorgestellten Ansätze zur ereignisbasierten Regelung physikalisch gekoppelter Systeme werden in Simulationen und Experimenten an einem thermofluiden Prozess erprobt (Abschnitte 3.4, 5.3, 5.4.6, 6.2.6, 6.4, 7.4). Diese Untersuchungen zeigen, dass

- die vorgestellten Methoden robust gegenüber Modellunsicherheiten sind,
- die Informationsrückkopplung durch die ereignisbasierte Regelung an das Systemverhalten angepasst wird und
- der Kommunikationsaufwand über die Rückführung gegenüber einer konventionellen zeitdiskreten Regelung deutlich reduziert wird.

1 Introduction

1.1 Event-based control

1.1.1 Structure

Event-based control is a means to restrict the feedback communication in a control loop to only those time instants, at which an event indicates the need for an update of the control input in order to satisfy a given specification of the closed-loop behavior. The event times are determined in dependence upon the states or the output of the system and in this way, the feedback communication is adapted to the system behavior. Since events are generally triggered asynchronously in time, the event-based control principle differs fundamentally from a conventional discrete-time control, where the feedback link is closed periodically.

> In sampled-data control the sampling and the update of the control input occur at equidistant time instants. The assumption of a fixed sampling period is violated in event-based control, where the control loop is closed only in response to the system behavior.

This consideration implies that the well-established theory on sampled-data control cannot be applied to event-based control. New methods for the design and analysis of event-based control must be developed instead, which explicitly take account of the fact that the control loop is closed only at the event times that occur non-periodically in time and is controlled in open loop in between consecutive events.

The structure of an event-based control loop for systems with N sensor nodes and N actuator nodes is shown in Fig. 1.1. It consists of the following components:

- The plant Σ with the inputs $\boldsymbol{u}_1(t), \ldots, \boldsymbol{u}_N(t)$, the states $\boldsymbol{x}_1(t), \ldots, \boldsymbol{x}_N(t)$ and the disturbance $\boldsymbol{d}(t)$.
- The event generators $E_1, \ldots, E_N$. The event generator E_i, $(i \in \mathcal{N} = \{1, \ldots, N\})$ determines the event time t_k at which current state information $\boldsymbol{x}_i(t_k)$ is transmitted over the communication network to the controllers. Moreover, it specifies to which control unit this information is transmitted.

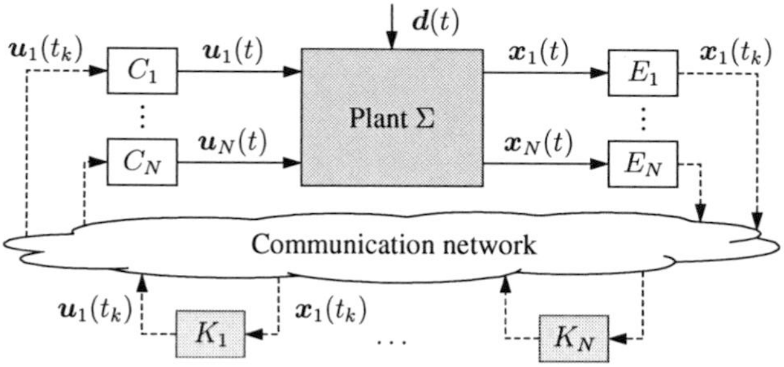

Figure 1.1: Structure of the event-based control loop

- The controller K consists of the decentralized control units $K_1, \ldots, K_N$. The control unit K_i $(i \in \mathcal{N})$ computes the control input $\boldsymbol{u}_i(t_k)$ using the information received at the event time t_k.

- The control input generators $C_1, \ldots, C_N$. The control input generator C_i $(i \in \mathcal{N})$ generates the control input $\boldsymbol{u}_i(t)$ based on the information $\boldsymbol{u}_i(t_k)$ for the time $t \geq t_k$ until the next event occurs.

In Fig. 1.1, the dashed lines symbolize the feedback communication that occurs at the event times only, whereas the solid lines indicate continuous information links.

1.1.2 Application fields

Event-based control is a paradigm where the controller reacts to only significant changes of the system behavior. Hence, the control task, which consists of the sampling of the system output, the recalculation of the control input and the updating of the actuator signals, is carried out when needed which is generally in an aperiodic manner. There are basically two motivations for using event-based control instead of conventional discrete-time control:

1. Changes in the measurement of the system output occur non-periodically and the sampling should be adapted to the behavior of the system.

2. The executions of the control task are costly in some sense and should be kept to a minimum.

These two motives are explained in the following by giving some application examples.

Control with asynchronous measurements. Encoders are generally event-based sensors [29], because the measurement is a quantization of the continuous signal. Sensors with AD converters have a limited resolution and if the resolution is low compared to the range of variation of the system output, changes of the measurement occur asynchronously in time which implies that a periodic execution of the control task is inappropriate. This problem was the starting point in [76] for the development of an event-based controller for the motor synchronization in a mailing system.

Other application examples are control systems where the controller should only react to a particular state of the system indicated by the measurement reaching or exceeding certain thresholds. In engine control the sampling is based on the angle position of the crankshaft rather than on time [29]. Similarly, in the control of rolling mills the sampling is based on flatness [89]. In [98] the Large Hadron Collider (LHC) at the European Organization for Nuclear Research (CERN) has been given as another application example for event-based control, because both the measurement and the control (acceleration and path keeping) of particles is based on their position within the accelerator ring. The event-based character of the control of this system is due to the fact that the trajectory of the particles can be manipulated only at some positions within the ring.

Control with minimum executions of the control task. Networked control systems are control loops where the feedback is closed over a real time communication network [86, 94, 154]. The execution of a control task in these systems requires the transmission of the current measurements to the controller and the transmission of the control input to the actuators. Hence, the feedback in network control systems increases the load of the communication network. By using event-based control methods the feedback communication effort and by implication the probability of delays or packet losses should be reduced compared to a discrete-time control.

The scheduling of control tasks on an embedded controller with limited computing capacity has served as the motivation for the development of an event-based control approach in [144]. The event-based scheduling aims at orchestrating the control tasks in the sense that the processing unit is allocated to a control task only when required in order to guarantee stability.

1.1.3 Fundamental questions

The investigation of event-based control grounds on the fundamental question, when the feedback link in a control loop needs to be closed in order to accomplish a given control task. Feedback is necessary in one of the following three situations [115]:

- An unstable plant needs to be stabilized.

- Feedback should allow the controller to deal with model uncertainties.
- Unknown disturbances need to be attenuated.

When considering control of physically interconnected systems, model uncertainties play a central role in the design and analysis of the controller. In this context the uncertainties do not refer to the system parameters but rather to the incompleteness of model information, because the design of the local control units for interconnected subsystems is usually based on partial model information of the overall system. Hence, the model uncertainties occur in the form of disregarded interconnections to other subsystems. These considerations lead to the following crucial questions for the design and analysis of the event-based controllers for interconnected systems:

- How should the components of the event-based controller be designed?
- At which time should information be transmitted?
- Which information should be transmitted at the event times?
- Between which components of the overall event-based controller is the exchange of information necessary?
- How should the received information be used by the components of the event-based controller?

These questions are the main concern of this thesis.

1.2 Literature on event-based control

1.2.1 Literature survey

The main characteristic of event-based control is the non-periodicity in the sampling of the output of the plant and the updating of the control signals. The interest in this control paradigm has grown significantly in the last decade [78], but the idea of aperiodic control as opposed to periodic control is not new and has been discussed in literature since controllers have been implemented on digital hardware. Under the catchword *adaptive sampling*, the variation of the sampling frequency in digital control loops has been investigated as a means to increase the efficiency of the sampling [33, 43, 73, 88, 124].

Although aperiodic control has been studied for a long time and all approaches in this research field have the same primary objective in common, namely the restriction of the sampling to only those times when it is necessary, there is no consistent terminology describing this control paradigm [98]. Some of the names that have been used in literature are minimum attention control [40], Lebesgue control [31], interrupt-based control [87], intermittent control [66], level-triggered control [134], level-crossing control [95], send-on-delta control [123] or deadband control [131]. The terms that can be predominately found in the current literature are event-based control [29], event-triggered control [106] and self-triggered control [159]. In this thesis the expression *event-based control* is primarily used.

Event-based control has the capability to reduce the communication over the feedback link, compared to discrete-time control, while accomplishing a desired control performance which has been shown in several simulation and experimental studies for a variety of different systems. The event-based control of electrical motors have been developed and evaluated for a mailing system and a printing system in [76] or [138], respectively, where the event-based sampling is motivated by the usage of low-resolution sensors that update of measurements asynchronously in time. The example of the control of DC-motors has also been used in [81, 172] in different applications to demonstrate the reduction of communication in event-based control systems. The works [68, 96, 100] have evaluated event-based control schemes for the examples of a chemical or a thermofluid process, showing that given performance requirements can be satisfied by using event-based control. Moreover, there are also examples which illustrate that the application of event-based sampling can lead to undesirable effects in the closed-loop system and, thus, requires a through investigation of its theoretical foundations. The work [42] has investigated the control of a solar collector field where the event-based sampling leads to limit cycles in case of disturbances. In [80] event-based sampling in engine control has been identified as a source of aliasing problems that can deteriorate the control performance.

Besides the investigation of the practical applicability, a lot of effort has been spent on developing a profound theory on event-based control. By emphasizing the benefits of event-based control in contrast to discrete-time control, the works [28, 30] have initiated the investigation of several design and analysis methods, like [31, 131, 169, 172]. In recent years, this research area has continuously grown and the theory on event-based control has been developed further in many different directions. Some of the most relevant lines of research on event-based control are discussed in the following.

Event-based state-feedback control. A large number of publications on event-based control is concerned with state-feedback control, like [77, 115, 131, 144, 166]. All cited approaches are called *emulation-based*, which means that the controller is designed for a conventional continuous or discrete-time state feedback and the design of an appropriate event triggering mechanism has to ensure the stability and desired performance of the event-based control system. The main difference in the previously cited works consist in the working principle of the triggering mechanism. In [77], events are triggered whenever the state exceeds a fixed threshold. Events are generated in [131], when the absolute difference between the plant state and previously sampled state are larger than a threshold and in [144, 166], when the difference between the state and the last sampled state relative to the current state violates a threshold. In [115], the triggering of events is caused when the plant state deviates from the prediction of the state by more than a predefined threshold. Whether the control system is asymptotically or practically stable depends upon the applied triggering mechanism [39].

References [37, 103] have presented approaches to *event-based PI-control*, which also require the plant state to be measurable. A different approach to event-based state feedback has been investigated in [83], called *sporadic control*, where at the event times the state is reset to zero using an impulsive control input. The works [44, 128] have presented optimization-based approaches to the joint design of the controller and the event triggering mechanism. In [137] it is explained that an optimal controller and triggering condition is hard to find, due to the so-called *dual effect* that describes the impact of the control input on the event triggering. Under certain conditions the presence of the triggering mechanism can be ignored and an optimal controller can be found.

The works [117, 119, 120, 132] have proposed event-based control approaches for systems with multiple sensors nodes. In these kind of systems synchronous transmissions from all sensors to the controller are undesirable which has motivated the investigation of decentralized triggering mechanisms for transmissions from the sensors to the controller.

Event-based output-feedback control. A severe restriction of the previously discussed event-based state-feedback approaches is that the state is required to be measurable. In many technical systems this presumption is not fulfilled, which has stimulated the development of *event-based output-feedback* approaches. References [101, 146, 150] and [107] have extended event-based state-feedback approaches to the output-feedback case by incorporating a state observer or a Kalman filter, respectively, in the event-based state-feedback controller. The state estimation, instead of the actual state, is then used for determining the event times and generating the control input.

There are only a few approaches in literature that tackle the problem of event-based output-feedback by not relying on state estimations. In [54, 62, 95, 125], the event triggering as well as the input generation is based on output measurements solely. The work [54] distinguishes from the remaining approaches in being the only one that has provided an analysis for the minimum inter-event time. Moreover, [54] also has considered control systems with distributed sensors and actuators and it has proposed a decentralized triggering mechanism for transmissions from sensors to the controller and from the controller to the actuators.

Self-triggered control and minimum attention control. As opposed to event-based control, *self-triggered control* is a proactive control scheme where the next sampling time is determined in advance based on the information that is available at the latest event [159]. In [60, 121] self-triggered control approaches have been presented for linear systems and in [23, 27, 118, 152] for nonlinear systems. Reference [25] has presented a method for the self-triggered implementation of a given event-based state feedback controller for nonlinear homogenous systems. In most approaches to self-triggered control the predetermination of the next sampling time is done under worst-case assumptions on the system behavior which results in the triggering of more events compared to event-based control. The conservatism in the calculation of the next event time is increased if systems with exogenous disturbances are considered [21, 153, 162, 164]. Self-triggered techniques have been implemented in [130] for the coordination of a network of robots.

The work [40] has introduced the idea of *minimum attention control*, considering that the cost for the implementation of a control law is linked to the variation of the state and the control input. The higher the rate is at which these signals change, the more attention requires the control law. Minimum attention control aims at minimizing the required attention in a control loop and, in this way, is closely related to self-triggered control. References [24, 55] have proposed minimum attention control approaches, where the control law is designed jointly with the triggering mechanism.

Comparison of event-based and discrete-time control. The first systematic methods for the comparison of event-based and discrete-time control have been published in [30, 31]. By considering first-order linear stochastic systems, these works have highlighted some advantages of event-based control over periodic control. These approaches have been pursued in [83] for the same class of systems, showing that event-based control can yield a better performance than periodic control with respect to the state variance and the reduction of the control effort. [104] has compared event-based and discrete-time control for linear systems of arbitrary order, proving that event-based control outperforms discrete-time control only under certain conditions.

Periodic event-triggered control. In *periodic event-triggered control* the triggering condition is checked periodically, rather than continuously as in most existing event-based control approaches. The basic idea of this control scheme is to combine the benefits of both event-based and discrete-time control. An advantage of verifying the event condition periodically is that the existence of a minimum inter-event time is inherently guaranteed. References [26, 75, 79, 160] have presented periodic event-triggered control approaches for linear systems, where [26] only has considered state-feedback control whereas [75, 79, 160] also have investigated output-feedback control laws. A decentralized control scheme has been discussed in

[75]. An extension to periodic event-triggered control of nonlinear systems has been presented in [133].

Event-based estimation. In literature, the use of event-based sampling techniques is not restricted to control but is also applied for state estimation. First approaches to *event-based state estimation* have been published in [45, 170]. The problem of designing appropriate triggering mechanisms for the event-based estimation has been considered in [127, 136] for first-order linear stochastic systems, in [74, 140] for linear systems and in [97] for nonlinear stochastic systems. Distributed event-based state estimation has been investigated in [155, 156, 167] and [168] has proposed a method for the event-based estimation over multi-hop networks.

Event-based model predictive control. The concept of event-based sampling has also stimulated the extension of classical control methods, like model predictive control (MPC). In *event-based MPC* the optimal control input sequence is not re-calculated at every time step, as in the conventional MPC, but only at those time steps where an event indicates the need for an update. Hence, the control signals of the latest sequence are implemented until the next event occurs. Since the iterations of the re-calculation of the control inputs are not carried out at every time step, event-based MPC reduces the computation effort compared to the conventional MPC. Event-based MPC for continuous-time nonlinear systems has been investigated in [56, 158] considering exogenous disturbances and in [157] concentrating on delays in the feedback. The coordination of multiple agents with nonlinear dynamics using event-based MPC has been studied in [57]. References [35, 105] have considered the event-based MPC of constrained linear systems. Approaches to self-triggered MPC have been developed in [34, 84].

Event-based control over networks. The main application field for event-based control methods are networked control systems. In these control systems the feedback link of the control loop is realized over a non-deterministic and unreliable real-time network [86, 94, 154]. Numerous publications in the literature on event-based control have been dedicated to the analysis of feedback communication imperfections or to the design of appropriate methods for the compensation of different effects of unreliable communication.

Event-based control with *delays* or *packet losses* in the feedback link has been investigated in [51, 62, 63, 71, 98, 102, 163, 174]. In [62, 63, 163, 174] the triggering mechanism has been adapted, taking the maximum network induced delay into account, so as to ensure that the closed-loop system remains stable. Reference [71] has proposed special communication protocols which include the transmission of acknowledge signals in order to compensate for severe delays or packet losses. Another compensation-based approach has been introduced in [98, 102], where the transmitted information is augmented with timestamps, which allows the receiver to identify the transmission delay. This information is then used to recover the trajectory of the control input that would have been applied in case of no delays.

The effect of *quantized state feedback* information has been studied in [63, 99, 147, 174], showing that closed-loop stability is retained by an appropriate design of the triggering mechanism. References [108, 109] have determined minimum bit-rates that are required to stabilize the event-based control system. In [69] an event-based control approach has been presented which is based on the quantization of the state space. In this work the event-based character lies in the asynchronous update of the control input only when the state enters a new quantization region.

References [46, 65] have presented resource-aware event-based control approaches for sensor-actuator networks. In contrast to most existing event-based control methods, the triggering mechanism in these works does not only decide whether to transmit current information or not, but also dictates the transmitters power level.

The works [38, 41, 82, 135] have investigated the interplay of multiple event-based control loops which are closed over a *shared communication medium.* In such networked control systems the access to the communication medium is restricted to a limited number of participants at a time and the rules for the allocation of the network access are defined by a communication protocol. This implies that the event-based control loops are interconnected over the common use of the shared communication medium. The works [38, 41, 82] have analyzed the performance of the event-based control loops in dependence upon different network access methods. Reference [135] has investigated the impact of packet losses on the performance of the event-based control loops, assuming that the loss rate is independent of the parameters of the triggering condition. An approach to adaptive event-based control has been presented in [126] where the event-based controllers adjust their transmission rate in order to meet global resource constraints. Reference [116] has presented a prioritized event-based scheduling policy which guarantees the stability of the overall control system.

Event-based control of multi-agent systems. Ideas from event-based sampling have also been applied to the coordination of multi-agent systems. Most works consider the event-based control of single integrator systems [52, 59, 64, 122] or double-integrator systems [139, 171]. The control of multi-agent systems is investigated in [49, 111] for agents with identical linear dynamics and in [173] for agents with nonlinear dynamics. Reference [91] has studied the problem of clock synchronization. In multi-agent systems the controlled agents are interconnected over a common control aim, like consensus [49, 122, 139, 171], synchronization [50, 91] or the shaping of a desired formation [59, 173], whereas the individual agents are not physically coupled.

Event-based control of interconnected systems. The problem of event-based control of system that are composed of physically interconnected subsystem has been tackled by only a few publications. Several works, like [54, 117, 119, 120, 132, 148], have proposed decentralized triggering mechanisms for the communication from the sensors to the controller or from the controller to the actuators of the plant, but none of these references has considered a decentralized controller structure. However, for the control of physically interconnected systems (which are often considered to be large-scale systems) a centralized controller is generally undesired.

The *distributed implementation of centralized control laws* using a decentralized architecture of the controller has been investigated by the works [61, 63, 172]. In these approaches the local control units predict the state of the overall control system and this prediction is applied to generate the control input for each subsystem. These predictions are updated at the event times using broadcast communication from one local controller to all other control units. The event-based control of the special class of hierarchically coupled systems has been investigated in [92].

Decentralized event-based control of interconnected systems has been considered in [75] in the framework of periodic event-triggered control. In this approach the control units are designed in a decentralized way and the stability of the overall control system is guaranteed by a condition that requires the subsystems to be weakly interconnected.

Distributed event-based control approaches have been proposed in [47, 161, 165] for nonlinear systems and in [48, 70] for linear systems. In this context distributed event-based control refers to control laws where the control input of the subsystems is a function of the local state and the state of several (but not all) other subsystems in the network. Extensions of these approaches to imperfect communication between the control units considering delays and packet losses has been studied in [71] for linear systems and in [163] for nonlinear systems. In all these references the stability of the overall system is ensured using small-gain arguments that claim the subsystems to be weakly connected in some sense.

Except for the reference [172], no approaches in the literature on event-based control of physically interconnected systems investigates the performance of the event-based control with respect to the capability of approximating a continuous control with adjustable accuracy. The approaches [61, 63, 172] that investigate distributed implementations of centralized control laws presume the state to be measurable. All decentralized and distributed event-based control approaches require the subsystems to be weakly coupled, which is a severe restriction on the system to be controlled. Moreover, these approaches only consider proactive transmissions of information, whereas they do not take information requests into account. The only approach to decentralized event-based control [75] studies periodic event-based control which is a special direction in the research area of event-based control and which differs from the investigations in this thesis.

1.2.2 Classification of this thesis

This thesis deals with the event-based state-feedback control of systems that are composed of physically interconnected subsystems. The subsystems are considered to have linear dynamics. The main control aim is the attenuation of exogenous disturbances, which are unknown but assumed to be bounded in magnitude. Hence, the overall control system shall be rendered practically stable with respect to a set (see Sec. 2.3). The proposed event-based control method relies on a decentralized controller structure, where each subsystem is controlled by a local control unit. The overall event-based controller encompasses the local control units and a communication network that is used for exchanging information between the control units. The network is assumed to be ideal, such that no delays, collisions or packet losses occur. Following the idea of [115], the proposed event-based controllers shall be made to approximate the behavior of a continuous state feedback with adjustable accuracy.

The event-based control methods that are proposed in this thesis are model-based approaches. This means that the event-based controller runs a dynamic model of the controlled plant in order to determine a prediction of the plant state which is used for generating the control input and for determining the event times. This principle differs from many other event-based control approaches which use a zero-order hold and do not further process the received information for generating the control input or determining the event times. Other model-based approaches to event-based control have been investigated in [58, 63, 75, 107, 115, 129, 172].

1.3 Contributions of the thesis

The contribution of this thesis is first discussed on an abstract level by explaining that the proposed event-based control methods are to be viewed as a part of a controller that aims at

- driving the plant state into a target region and
- keeping the state within that target region despite exogenous disturbances or the interaction of the subsystems.

Because the further investigations concern the second step, this discussion also justifies the standing assumption that the plant dynamics are accurately described by linear models. Afterwards, the contribution of this thesis with respect to the development of a theory on event-based control of interconnected systems is stated.

1.3.1 Global and local approach to event-based control

The event-based state-feedback approaches that are proposed in this thesis rely on a linear state-space model of the overall plant. The behavior of most technical systems, however, is described

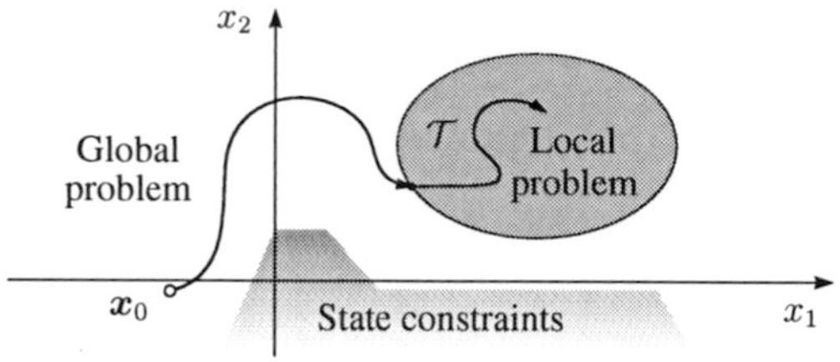

Figure 1.2: The global and the local problem

by a nonlinear model from which a linear model can be obtained by means of linearization in some point of the state-space, subsequently referred to as the *operating point*. In general, the linearized model and, by implication the analysis methods for the event-based control approaches, are only valid in a bounded region around this operating point.

The assumption that the plant is adequately described by a linear state-space model is justified by the control aim that all event-based control approaches proposed in this thesis concern disturbance rejection. Hence, the plant state $\boldsymbol{x}(t)$ should be kept within some target region $\mathcal{T} \subset \mathbb{R}^n$ (where the linearized model is valid) by the event-based controller in spite of exogenous disturbances and physical interconnections between the subsystems.

In general the initial plant state $\boldsymbol{x}_0$ cannot be expected to be sufficiently close to the operating point. In that case, the problem of stabilizing the plant in the set $\mathcal{T}$ can be subdivided into a global problem and a local problem (Fig. 1.2). The references [1, 68] have proposed to solve these problems by the following two complementary approaches to event-based control.

1. The *global event-based control approach* drives the plant state $\boldsymbol{x}(t)$ from the initial state $\boldsymbol{x}_0$ into the target region $\mathcal{T}$ while taking possible constraints on the states or control input into account. The global approach has to consider the nonlinear model of the overall plant, since the distance in the state-space between the initial state $\boldsymbol{x}_0$ and the target set $\mathcal{T}$ can be arbitrarily large.

2. The *local event-based control approach* makes the set $\mathcal{T}$ robustly positive invariant, i.e., once $\boldsymbol{x}(t)$ has entered $\mathcal{T}$ it should be kept within this set in spite of exogenous disturbances and interconnections between the subsystems.

Following this idea, the overall event-based controller works in two modes. As long as the state $\boldsymbol{x}(t)$ has not reached the target set $\mathcal{T}$ the controller operates according to the global event-based control approach. Several event-based control approaches have been published in literature which could be used as the global approach, such as [1, 110]. At the time $t = T_s$ when the state $\boldsymbol{x}(t)$ enters the target set $\mathcal{T}$ the controller switches to the local event-based control approach which from then on keeps the state $\boldsymbol{x}(t)$ in $\mathcal{T}$.

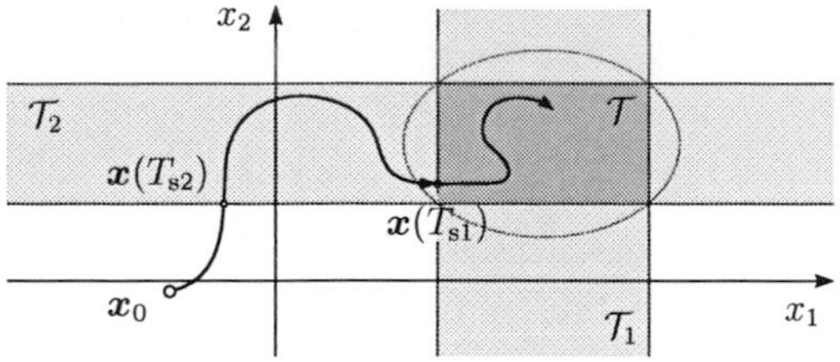

Figure 1.3: Switching from the global to the local approach in different subsystems

> The methods for event-based state feedback that are elaborated in this thesis are applicable as local event-based control approaches.

Since the event-based control of interconnected systems is investigated in this thesis the local and the global problem are reformulated from a subsystem perspective. Consider that the target region can be stated as

$$\mathcal{T} \supseteq \bigcap_{i \in \mathcal{N}} \mathcal{T}_i$$

where $\mathcal{T}_i$ denotes the target region for the state $\boldsymbol{x}_i$ of subsystem Σ_i and note that $\boldsymbol{x}_i \in \mathcal{T}_i$ for all $i \in \mathcal{N} = \{1, \ldots, N\}$ implies $\boldsymbol{x} \in \mathcal{T}$. Hence, the global event-based control approach for subsystem Σ_i aims at steering the subsystem state $\boldsymbol{x}_i(t)$ from the initial state $\boldsymbol{x}_{0i}$ to the target region $\mathcal{T}_i$, where at time $t = T_{si}$ the local event-based control approach takes over and keeps the state $\boldsymbol{x}_i(t)$ in $\mathcal{T}_i$ in spite of the local disturbance and interconnections to other subsystems. The switching from the global to the local approach for different subsystems can occur at different times, i.e., generally $T_{si} \neq T_{sj}$ for arbitrary $i, j \in \mathcal{N}$, $i \neq j$. Thus, the switching to the local approach occurs asynchronously in time which is illustrated in Fig. 1.3.

The asynchronous activation of the local approaches for the several subsystems is not explicitly considered in this thesis, as it mainly concerns the implementation of the event-based state-feedback controllers, but does not affect their design or analysis.

1.3.2 Theoretical contributions

The main contribution of this thesis consists in the extension of the event-based state-feedback approach [115] from systems that have centralized sensor and actuator nodes to systems which are composed of physically interconnected systems. The focus is on the investigation of three different structures for event-based state-feedback control of interconnected systems which basically differ with respect to

- the *model information* used in the components of the event-based controller,
- the *communication effort* for information transmissions between the controller components and
- the *applicability* to interconnected systems in dependence upon the coupling strength between the subsystems.

Structures. The underlying idea of all approaches is that the event-based state-feedback controller shall be designed such that it approximates the behavior of a continuous state feedback with adjustable accuracy. The investigation of the feasibility of this design aim leads to a condition on the structure of the implemented state-feedback gain and on the communication topology for the information transmissions between the components of the controller (Theorem 4.1). This condition can be satisfied in various ways which is exploited in the elaboration of the three proposed event-based controller structures, which are briefly outlined in the following.

The **distributed realization of the centralized event-based state-feedback approach** [115] (Chapter 5) aims at achieving the same control performance which would be obtained by a centralized event-based state feedback. The model information of the overall plant is used in each component of the controller and at the event times the information is broadcasted from one control unit to all the others. The approach is applicable to interconnected systems with arbitrary coupling strength.

In the approach to **decentralized event-based state-feedback control** (Chapter 6) each component of the overall controller includes local model information only and information is solely communicated locally from the event generator to the corresponding control input generator of the respective controller component. The applicability of this approach is restricted to systems which are weakly coupled in a well defined sense.

In the **distributed event-based state-feedback control** (Chapter 7) the approximation of the behavior of the continuous state feedback is accomplished by using a new kind of event-based control where two types of event conditions trigger both the transmission of current state information as well as the request of information from neighboring subsystems. This approach allows for an arbitrary coupling strength between neighboring systems but requires the coupling strength between subsystems that are not directly interconnected to be weak.

Results. For all these approaches, it is shown that

- the event-based control system approximates the behavior of a continuous state-feedback loop with adjustable accuracy (Theorems 5.1, 6.1, 7.3),

- the state of the plant is maintained by the event-based state feedback in a bounded surrounding of the origin, the size of which depends upon the event threshold parameters of the control units (Theorem 6.4, Corollary 7.1) and
- the minimum time that elapses in between two consecutive events is bounded from below (Theorems 5.2, 6.6, 7.4, 7.5).

Additional results. Moreover, the thesis presents

- new methods for the stability analysis of interconnected event-based state-feedback loops (Theorem 6.2 for decentralized event-based control and Theorems 7.1, 7.3 for distributed event-based control) using the comparison principle [113],
- two methods for the estimation of coupling input signals (Section 6.2.5) which can be used to reduce the triggering of events in decentralized event-based control,
- a method for the event-based control of interconnected systems, where the state of some subsystems is not accessible for measurement (Section 5.4).

Experimental evaluations. All proposed event-based control approaches are evaluated in simulations and experiments on a thermofluid process (Sections 3.4, 5.3, 5.4.6, 6.2.6, 6.4, 7.4). The results of these evaluations indicate that

- the proposed event-based control methods are robust with respect to model uncertainties,
- the communication over the feedback link is adapted to the system behavior and
- the feedback communication effort is considerably reduced compared to a conventional discrete-time control.

1.4 Structure of this thesis

Chapter 2 introduces the notation and some basic definitions that are used throughout the thesis. The models for the subsystems, their interconnection and the communication network are introduced. Finally, a thermofluid process that consists of two interconnected subsystems is described. This process is used to illustrate the proposed control methods and to evaluate the analysis methods by means of simulations and examples.

Chapter 3 summarizes the event-based state feedback approach [115] and states its main properties.

Chapter 4 introduces the structure of the event-based controller for interconnected systems that is the basic framework in which the event-based control methods presented in the remaining chapters are developed. A general condition for the boundedness of the deviation between the behavior of the event-based state-feedback system and the continuous state-feedback loop is derived. This condition, which can be satisfied in various ways, represents the foundation for the development of the event-based control methods proposed in this thesis.

Chapter 5 presents a method for the event-based control of interconnected systems that is a distributed implementation of the centralized event-based state-feedback approach [115]. The conceptual differences and similarities between both approaches are investigated by analyzing the approximation error between the event-based and the continuous state-feedback system as well as the minimum inter-event time. Moreover, the proposed approach is extended to the case where the state in some subsystems is not accessible for measurement.

Chapter 6 investigates the decentralized event-based control of physically interconnected systems. First, a method for the event-based implementation of a given decentralized state-feedback law is presented. Two methods for the coupling signal estimation are proposed with the aim of reducing the triggering of events. Second, the stability of interconnected event-based state-feedback loops is analyzed where the event-based controllers are designed following the approach [115], neglecting the impact of the interconnections.

Chapter 7 first presents a new method for the design of a distributed state-feedback law that ensures the stability of the overall control system with continuous state feedback. Afterwards, a method for the implementation of the distributed state feedback in an event-based manner is proposed which leads to a new kind of event-based control, where events do not only trigger the transmission of information but also the request of information from neighboring subsystems.

Chapter 8 summarizes and concludes the thesis. Possible directions for further research are discussed.

2 Preliminaries

This chapter starts with the introduction of the notation that will be used consistently throughout the thesis. The models for subsystems and their interconnections are given, followed by a description of the considered communication network. The concept of practical stability *is formulated as the notion of stability applied in this thesis. Finally, a thermofluid process is introduced which serves as a demonstration example for the presented event-based control concepts.*

2.1 Notation and definitions

2.1.1 General notations

$\mathbb{R}$ and $\mathbb{R}_+$ denote the set of real numbers or the set of positive real numbers, respectively. $\mathbb{N}$ is the set of natural numbers and $\mathbb{N}_0 = \mathbb{N} \cup \{0\}$. Throughout this thesis scalars are denoted by italic letters ($s \in \mathbb{R}$), vectors by bold italic letters ($\boldsymbol{x} \in \mathbb{R}^n$) and matrices by upper-case bold italic letters ($\boldsymbol{A} \in \mathbb{R}^{n \times n}$).

If $\boldsymbol{x}$ is a signal, the value of $\boldsymbol{x}$ at time $t \in \mathbb{R}_+$ is represented by $\boldsymbol{x}(t)$. Moreover,

$$\boldsymbol{x}(t^+) = \lim_{s \downarrow t} \boldsymbol{x}(s)$$

represents the limit of $\boldsymbol{x}(t)$ taken from above.

The transpose of a vector $\boldsymbol{x}$ or a matrix $\boldsymbol{A}$ is denoted by $\boldsymbol{x}^\top$ or $\boldsymbol{A}^\top$, respectively. $\boldsymbol{I}_n$ denotes the identity matrix of size n and $\mathbf{1}_n$ is an n-dimensional vector where each element is 1. $\boldsymbol{O}_{n \times m}$ is the zero matrix with n rows and m columns and $\mathbf{0}_n$ represents the zero vector of dimension n. The dimensions of these matrices and vectors are omitted if they are clear from the context.

Consider the matrices $\boldsymbol{A}_1, \ldots, \boldsymbol{A}_N$. The notation $\boldsymbol{A} = \operatorname{diag}(\boldsymbol{A}_1, \ldots, \boldsymbol{A}_N)$ is used to denote a block diagonal matrix

$$\boldsymbol{A} = \operatorname{diag}\left(\boldsymbol{A}_1, \ldots, \boldsymbol{A}_N\right) = \begin{pmatrix} \boldsymbol{A}_1 & & \\ & \ddots & \\ & & \boldsymbol{A}_N \end{pmatrix}.$$

The i-th eigenvalue of a square matrix $\boldsymbol{A} \in \mathbb{R}^{n \times n}$ is denoted by $\lambda_i(\boldsymbol{A})$. The matrix $\boldsymbol{A}$ is called Hurwitz (or stable) if $\operatorname{Re}\big(\lambda_i(\boldsymbol{A})\big) < 0$ holds for all $i = 1, \ldots, n$, where $\operatorname{Re}(\cdot)$ denotes the real part of the indicated number.

Consider a time-dependent matrix $\boldsymbol{G}(t)$ and vector $\boldsymbol{u}(t)$. The asterisk $*$ is used to denote the convolution-operator, e.g.,

$$\boldsymbol{G} * \boldsymbol{u} = \int_0^t \boldsymbol{G}(t-\tau)\boldsymbol{u}(\tau)\mathrm{d}\tau.$$

The inverse of a square matrix $\boldsymbol{H} \in \mathbb{R}^{n \times n}$ is symbolized by $\boldsymbol{H}^{-1}$. For a non-square matrix $\boldsymbol{H}^{n \times m}$ that has full rank, $\boldsymbol{H}^{+}$ is the pseudoinverse that is defined by

$$\boldsymbol{H}^{+} = \begin{cases} \left(\boldsymbol{H}^{\top}\boldsymbol{H}\right)^{-1}\boldsymbol{H}^{\top}, & \text{for } n > m \\ \boldsymbol{H}^{\top}\left(\boldsymbol{H}^{\top}\boldsymbol{H}\right)^{-1}, & \text{for } m > n \end{cases}$$

For $n > m$, $\boldsymbol{H}^{+}$ denotes the left inverse ($\boldsymbol{H}^{+}\boldsymbol{H} = \boldsymbol{I}_m$) and for $m > n$ $\boldsymbol{H}^{+}$ is the right inverse ($\boldsymbol{H}\boldsymbol{H}^{+} = \boldsymbol{I}_n$) [36].

For two vectors $\boldsymbol{v}, \boldsymbol{w} \in \mathbb{R}^n$ the relation $\boldsymbol{v} > \boldsymbol{w}$ ($\boldsymbol{v} \geq \boldsymbol{w}$) holds element-wise, i.e., $v_i > w_i$ ($v_i \geq w_i$) is true for all $i = 1, \ldots, n$, where v_i and w_i refer to the i-th element of the vectors $\boldsymbol{v}$ or $\boldsymbol{w}$, respectively. Accordingly, for two matrices $\boldsymbol{V}, \boldsymbol{W} \in \mathbb{R}^{n \times m}$ where $\boldsymbol{V} = (v_{ij})$ and $\boldsymbol{W} = (w_{ij})$ are composed of the elements v_{ij} and w_{ij} for $i = 1, \ldots, n$ and $j = 1, \ldots m$ the relation $\boldsymbol{V} > \boldsymbol{W}$ ($\boldsymbol{V} \geq \boldsymbol{W}$) refers to $v_{ij} > w_{ij}$ ($v_{ij} \geq w_{ij}$). For a scalar s, $|s|$ denotes the absolute value. For a vector $\boldsymbol{x} \in \mathbb{R}^n$ or a matrix $\boldsymbol{A} = (a_{ij}) \in \mathbb{R}^{n \times m}$ the $|\cdot|$-operator holds element-wise, i.e.,

$$|\boldsymbol{x}| = \begin{pmatrix} |x_1| \\ \vdots \\ |x_n| \end{pmatrix}, \qquad |\boldsymbol{A}| = \begin{pmatrix} |a_{11}| & \ldots & |a_{1m}| \\ \vdots & \ddots & \vdots \\ |a_{n1}| & \ldots & |a_{nm}| \end{pmatrix}.$$

$\|\boldsymbol{x}\|$ and $\|\boldsymbol{A}\|$ denote an arbitrary vector norm and the induced matrix norm according to

$$\|\boldsymbol{x}\|_p := \left(\sum_{i=1}^{n} |x_i|^p\right)^{\frac{1}{p}}, \qquad \|\boldsymbol{A}\|_p := \max_{\boldsymbol{x}\neq \boldsymbol{0}} \frac{\|\boldsymbol{A}\boldsymbol{x}\|_p}{\|\boldsymbol{x}\|_p}$$

with the real number $p \geq 1$ and $\|\boldsymbol{x}\|_\infty$ refers to the uniform norm

$$\|\boldsymbol{x}\|_\infty := \max_{i\in\{1,\dots,n\}} |x_i|\,.$$

Sets are denoted by calligraphic letters ($\mathcal{A} \subset \mathbb{R}^n$). For the compact set $\mathcal{A} \subset \mathbb{R}^n$

$$\|\boldsymbol{x}\|_{\mathcal{A}} := \inf\left\{\|\boldsymbol{x} - \boldsymbol{z}\| \;\middle|\; \boldsymbol{z} \in \mathcal{A}\right\}$$

denotes the point-to-set distance from $\boldsymbol{x} \in \mathbb{R}^n$ to $\mathcal{A}$ [142].

2.1.2 Non-negative matrices and M-matrices

Definition 2.1 *A matrix* $\boldsymbol{A} = (a_{ij})$ *is called non-negative* ($\boldsymbol{A} \geq \boldsymbol{O}$) *if all elements of* $\boldsymbol{A}$ *are real and non-negative* ($a_{ij} \geq 0$).

Theorem 2.1 (Theorem A1.1 in [113], Perron-Frobenius theorem) *Every irreducible non-negative matrix* $\boldsymbol{A} \in \mathbb{R}_+^{n\times n}$ *has a positive eigenvalue* $\lambda_{\mathrm{P}}(\boldsymbol{A})$ *that is not exceeded by any other eigenvalue* $\lambda_i(\boldsymbol{A})$

$$\lambda_{\mathrm{P}}(\boldsymbol{A}) \geq |\lambda_i(\boldsymbol{A})|\,.$$

$\lambda_{\mathrm{P}}(\boldsymbol{A})$ is called the *Perron root* of $\boldsymbol{A}$.

Definition 2.2 *A matrix* $\boldsymbol{P} = (p_{ij}), \boldsymbol{P} \in \mathbb{R}^{n\times n}$ *is said to be an M-matrix if* $p_{ij} \leq 0$ *holds for all* $i \neq j$ *and all eigenvalues of* $\boldsymbol{P}$ *have positive real part.*

The following theorem presents an important property of M-matrices which is used in this thesis.

Theorem 2.2 (Theorem A1.3 in [113]) *A matrix* $\boldsymbol{P} = (p_{ij}), \boldsymbol{P} \in \mathbb{R}^{n\times n}$ *with* $p_{ij} \leq 0$ *for all* $i \neq j$ *is an M-matrix if and only if* $\boldsymbol{P}$ *is non-singular and* $\boldsymbol{P}^{-1}$ *is non-negative.*

The next theorem makes a relationship between non-negative matrices and M-matrices.

Theorem 2.3 (Theorem A1.5 in [113]) *If $\boldsymbol{A} \in \mathbb{R}_{+}^{n \times n}$ is a non-negative matrix, then*

$$\boldsymbol{P} = \mu \boldsymbol{I}_n - \boldsymbol{A}$$

with $\mu \in \mathbb{R}_{+}$ is an M-matrix if and only if

$$\mu > \lambda_{\mathrm{P}}(\boldsymbol{A})$$

holds.

2.2 Models

2.2.1 Plant

The overall plant is represented by the linear state-space model

$$\Sigma : \quad \dot{\boldsymbol{x}}(t) = \boldsymbol{A}\boldsymbol{x}(t) + \boldsymbol{B}\boldsymbol{u}(t) + \boldsymbol{E}\boldsymbol{d}(t), \quad \boldsymbol{x}(0) = \boldsymbol{x}_0 \tag{2.1}$$

where $\boldsymbol{x} \in \mathbb{R}^n$, $\boldsymbol{u} \in \mathbb{R}^m$ and $\boldsymbol{d} \in \mathcal{D}$ denote the state, the control input and the disturbance, respectively. Concerning the properties of the system (2.1) the following assumptions are made, unless otherwise stated:

A 1.1 The plant dynamics are accurately known.

A 1.2 The state $\boldsymbol{x}(t)$ is measurable.

A 1.3 The disturbance $\boldsymbol{d}(t)$ is bounded according to

$$\boldsymbol{d}(t) \in \mathcal{D} := \left\{ \boldsymbol{d} \in \mathbb{R}^p \middle| \, |\boldsymbol{d}| \leq \bar{\boldsymbol{d}} \right\}, \quad \forall \, t \geq 0 \tag{2.2}$$

where $\bar{\boldsymbol{d}} \in \mathbb{R}_{+}^{p}$ denotes the bound on the magnitude of the disturbance $\boldsymbol{d}$.

Equation (2.1) is referred to as the *unstructured model* [113], because it barely gives an insight into the interconnections and the dynamics of the subsystems which form the overall plant. The model (2.1) is useful if the overall system has a central sensor unit that collects all measurements and a centralized actuator unit that drives all actuators of the plant (Fig. 2.1(a)). The next section introduces a model which reflects the internal structure of the overall plant in more detail.

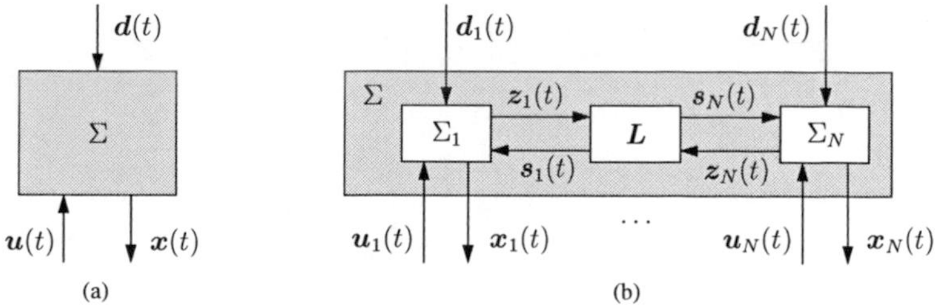

Figure 2.1: Structure of the overall system with (a) central sensor unit and actuator unit and (b) a sensor unit and actuator unit for each subsystem.

2.2.2 Interconnected subsystems

In this thesis the overall plant is considered to be composed of N physically interconnected subsystems. The subsystem $i \in \mathcal{N} = \{1, \dots, N\}$ is represented by the linear state-space model

$$\Sigma_i : \begin{cases} \dot{\boldsymbol{x}}_i(t) = \boldsymbol{A}_i \boldsymbol{x}_i(t) + \boldsymbol{B}_i \boldsymbol{u}_i(t) + \boldsymbol{E}_i \boldsymbol{d}_i(t) + \boldsymbol{E}_{\mathrm{s}i} \boldsymbol{s}_i(t), \quad \boldsymbol{x}_i(0) = \boldsymbol{x}_{0i} \\ \boldsymbol{z}_i(t) = \boldsymbol{C}_{\mathrm{z}i} \boldsymbol{x}_i(t) \end{cases} \tag{2.3}$$

where $\boldsymbol{x}_i \in \mathbb{R}^{n_i}$, $\boldsymbol{u}_i \in \mathbb{R}^{m_i}$, $\boldsymbol{d}_i \in \mathcal{D}_i$, $\boldsymbol{s}_i \in \mathbb{R}^{q_i}$ and $\boldsymbol{z}_i \in \mathbb{R}^{r_i}$ denote the state, the control input, the disturbance, the coupling input and the coupling output, respectively. The subsystem Σ_i is interconnected with the remaining subsystems according to the relation

$$\boldsymbol{s}_i(t) = \sum_{j=1}^{N} \boldsymbol{L}_{ij} \boldsymbol{z}_j(t) \tag{2.4}$$

where the matrix $\boldsymbol{L}_{ij} \in \mathbb{R}^{q_i \times r_j}$ represents the couplings from some subsystem Σ_j to subsystem Σ_i. The model (2.3), (2.4) is called the *interconnection-oriented model* [113]. Concerning this model the following assumptions are made:

A 1.4 The pair $(\boldsymbol{A}_i, \boldsymbol{B}_i)$ is controllable for each $i \in \mathcal{N}$.

A 1.5 $\boldsymbol{L}_{ii} = \boldsymbol{O}$ holds for all $i \in \mathcal{N}$, i.e., the coupling input $\boldsymbol{s}_i(t)$ does not directly depend upon the coupling output $\boldsymbol{z}_i(t)$.

The last assumption is weak and can always be fulfilled by modeling all internal dynamics in the matrix $\boldsymbol{A}_i$ for all $i \in \mathcal{N}$. Note that the assumptions **A** 1.1 to **A** 1.3 imply that

- the dynamics of the subsystems are accurately known,
- the subsystem state $\boldsymbol{x}_i(t)$ is measurable for each $i \in \mathcal{N}$ and

- the local disturbance $\boldsymbol{d}_i(t)$ is bounded according to

$$\boldsymbol{d}_i(t) \in \mathcal{D}_i := \left\{\boldsymbol{d}_i \in \mathbb{R}^{p_i} \middle| \, |\boldsymbol{d}_i| \leq \bar{\boldsymbol{d}}_i\right\}, \quad \forall\, t \geq 0 \tag{2.5}$$

  with $\bar{\boldsymbol{d}}_i \in \mathbb{R}_+^{p_i}$ representing the bound on the local disturbance $\boldsymbol{d}_i(t)$.

The unstructured model (2.1) can be determined from the interconnection-oriented model (2.3), (2.4) by taking

$$\boldsymbol{A} = \operatorname{diag}(\boldsymbol{A}_1, \ldots, \boldsymbol{A}_N) + \operatorname{diag}(\boldsymbol{E}_{\mathrm{s}1}, \ldots, \boldsymbol{E}_{\mathrm{s}N})\, \boldsymbol{L} \operatorname{diag}(\boldsymbol{C}_{\mathrm{z}1}, \ldots, \boldsymbol{C}_{\mathrm{z}N}), \tag{2.6a}$$

$$\boldsymbol{B} = \operatorname{diag}(\boldsymbol{B}_1, \ldots, \boldsymbol{B}_N), \tag{2.6b}$$

$$\boldsymbol{E} = \operatorname{diag}(\boldsymbol{E}_1, \ldots, \boldsymbol{E}_N) \tag{2.6c}$$

with the overall interconnection matrix

$$\boldsymbol{L} = \begin{pmatrix} \boldsymbol{O} & \boldsymbol{L}_{12} & \ldots & \boldsymbol{L}_{1N} \\ \boldsymbol{L}_{21} & \boldsymbol{O} & \ldots & \boldsymbol{L}_{2N} \\ \vdots & \vdots & \ddots & \vdots \\ \boldsymbol{L}_{N1} & \boldsymbol{L}_{N2} & \ldots & \boldsymbol{O} \end{pmatrix} \tag{2.7}$$

and by assembling the signal vectors according to

$$\boldsymbol{x}(t) = \begin{pmatrix} \boldsymbol{x}_1^\top(t) & \ldots & \boldsymbol{x}_N^\top(t) \end{pmatrix}^\top,$$
$$\boldsymbol{u}(t) = \begin{pmatrix} \boldsymbol{u}_1^\top(t) & \ldots & \boldsymbol{u}_N^\top(t) \end{pmatrix}^\top,$$
$$\boldsymbol{d}(t) = \begin{pmatrix} \boldsymbol{d}_1^\top(t) & \ldots & \boldsymbol{d}_N^\top(t) \end{pmatrix}^\top.$$

2.2.3 Communication network

Throughout this thesis the communication network is assumed to be ideal in the following sense:

A 1.6 The transmission of information over the communication network happens instantaneously, without delays and packet losses. Multiple transmitters can simultaneously send information without collisions occurring.

From an information theoretic point of view, this assumption is of course impossible to satisfy. However, in this thesis the communication network is considered to transmit information much faster compared to the dynamical behavior of the control system, which from a control theoretic perspective justifies the assumption **A** 1.6.

2.3 Practical stability

Event-based control systems are hybrid dynamical systems and, more specifically, belong to the class of impulsive systems [5, 54]. The notion of stability for these kinds of systems that is commonly used in literature is the stability with respect to compact sets [53, 67].

Definition 2.3 *Consider the system* (2.1) *and a compact set* $\mathcal{A} \subset \mathbb{R}^n$.

- *The set* $\mathcal{A}$ *is stable for the system* (2.1) *if for each* $\varepsilon > 0$ *there exists* $\delta > 0$ *such that* $\|\boldsymbol{x}(0)\|_{\mathcal{A}} \leq \delta$ *implies* $\|\boldsymbol{x}(t)\|_{\mathcal{A}} \leq \varepsilon$ *for all* $t \geq 0$.

- *The set* $\mathcal{A}$ *is globally attractive for the system* (2.1) *with* (2.2) *if*

$$\lim_{t \to \infty} \|\boldsymbol{x}(t)\|_{\mathcal{A}} = 0$$

  holds for all $\boldsymbol{x}(0) \in \mathbb{R}^n$.

- *The set* $\mathcal{A}$ *is globally asymptotically stable for the system* (2.1) *if it is stable and globally attractive.*

This definition claims that for any bounded initial condition $\boldsymbol{x}(0)$ the state trajectory $\boldsymbol{x}(t)$ remains bounded for all $t \geq 0$ and eventually converges to the set $\mathcal{A}$ for $t \to \infty$.

Definition 2.4 *If* $\mathcal{A}$ *is a globally asymptotically stable set for the system* (2.1)*, then* (2.1) *is said to be* practically stable with respect to the set $\mathcal{A}$.

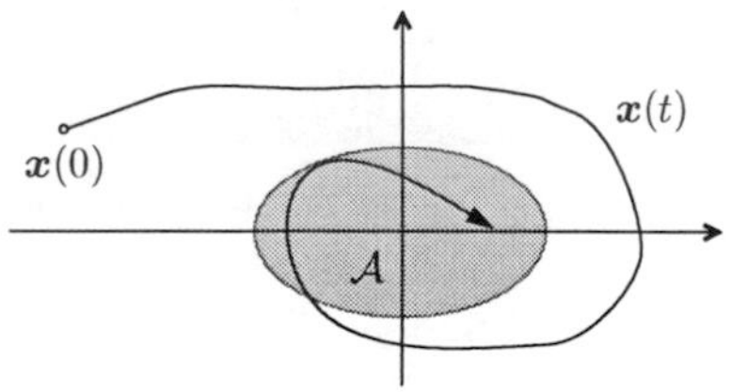

Figure 2.2: Practical stability with respect to the set $\mathcal{A}$

A graphical interpretation of this stability notion is given in Fig. 2.2, which shows the trajectory $\boldsymbol{x}(t)$ of a system of the form (2.1) in the two-dimensional state-space. The trajectory $\boldsymbol{x}(t)$ is bounded to the set $\mathcal{A}$ for $t \to \infty$. Note that the size of $\mathcal{A}$ depends upon the disturbance magnitude $\bar{d}$ and, in case of event-based state feedback, also on some parameters of the event-based controller, as will be explained in more detail later.

Relations to other notions of stability. The concept of practical stability of the system (2.1) is closely related to *input-to-state stability* (ISS) [141]. In [142] it has been shown that ISS with respect to some closed set $\mathcal{A}$ is equivalent to *input-to-state practical stability* (ISpS). ISpS has been used in the context of event-based control in [1].

Ultimate boundedness is another notion of stability that is also often used when investigating stability of event-based control systems [1, 98, 145], which is defined as follows [93]:

Definition 2.5 *The system* (2.1) *with* (2.2) *is said to be globally uniformly ultimately bounded if for every* $\boldsymbol{x}(0)$ *there exists* $p \in \mathbb{R}_+$ *and some time* T *such that*

$$\boldsymbol{x}(t) \in \Omega := \left\{ \boldsymbol{x} \in \mathbb{R}^n \middle| \, \|\boldsymbol{x}\| \leq p \right\}, \quad \forall \, t \geq T.$$

Note that there are subtle difference between the concepts of practical stability and ultimate boundedness. A system is ultimately bounded if the state $\boldsymbol{x}(t)$ enters some compact set Ω within a finite time T, whereas the concept of practical stability claims asymptotic convergence to a set $\mathcal{A}$. In this sense, ultimate boundedness is a stronger statement of stability, because it requires the state $\boldsymbol{x}(t)$ to actually reach the set Ω in finite time and not only asymptotically.

2.4 Demonstration example: Two interconnected thermofluid processes

The event-based control approaches presented in this thesis are tested and the analytical results are quantitatively evaluated through simulations and experiments on a demonstration process realized at the pilot plant at the Institute of Automation and Computer Control at Ruhr-University Bochum, Germany (Fig. 2.3). The plant includes four cylindrical storage tanks, three batch reactors and a buffer tank which are connected over a complex pipe system and it is constructed with standard industrial components including more than 70 sensors and 80 actuators.

Process description. The experimental setup of the process is illustrated in Fig. 2.4. The main components are the two reactors TB and TS used to realize continuous flow processes. Reactor TB is connected to the storage tank T_1 from where the inflow can be controlled by means of the valve angle $u_{\mathrm{T1}}(t)$. Via the pump PB a part of the outflow is pumped out into the buffer tank TW (and is not used further in the process) while the remaining outflow is conducted to the reactor TS. The temperature $\vartheta_{\mathrm{B}}(t)$ of the water in reactor TB is influenced by the cooling unit (CU) using the input $u_{\mathrm{CU}}(t)$ or by the heating rods that are driven by the signal $d_{\mathrm{H}}(t)$. The inflow from the storage tank T_3 to the reactor TS can be adjusted by means of the opening angle $u_{\mathrm{T3}}(t)$. Reactor TS is additionally fed by the fresh water supply (FW) from where the inflow is

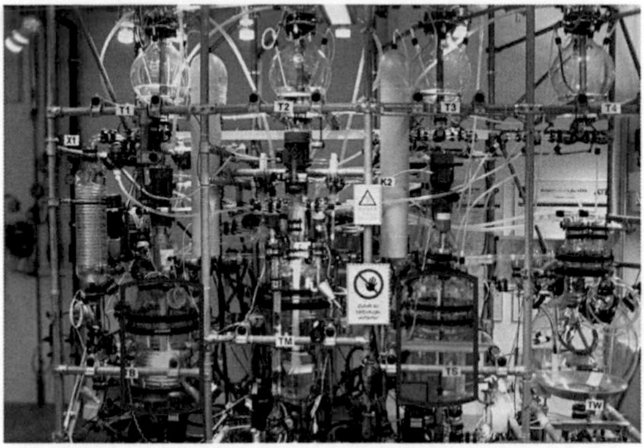

Reactor TB Reactor TS

Figure 2.3: Pilot plant, where the reactors which are used for the considered process are highlighted.

controlled by means of the valve angle $d_{\mathrm{F}}(t)$. Equivalently to reactor TB, the outflow of reactor TS is split and one part is conveyed via the pump PS to TW and the other part is pumped to the reactor TB. The temperature $\vartheta_{\mathrm{TS}}(t)$ of the liquid in reactor TS can be increased by the heating rods that are controlled by the signal $u_{\mathrm{H}}(t)$. The signals $d_{\mathrm{H}}(t)$ and $d_{\mathrm{F}}(t)$ are disturbance inputs which are used to define test scenarios with reproducible disturbance characteristics, whereas the signals $u_{\mathrm{T1}}(t)$, $u_{\mathrm{CU}}(t)$, $u_{\mathrm{T3}}(t)$ and $u_{\mathrm{H}}(t)$ are control inputs. Both reactor TB and reactor TS are equipped with sensors that continuously measure the level and the temperature of the contents.

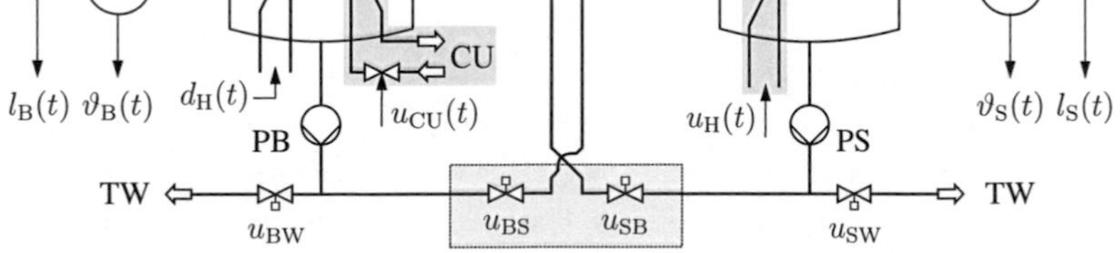

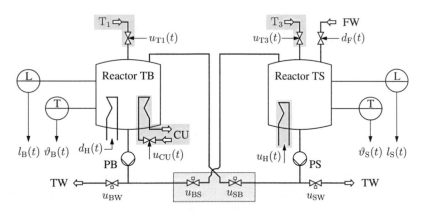

Figure 2.4: Experimental setup of the continuous flow process

The two reactors are coupled by the flow from reactor TB to reactor TS and vice versa, where the coupling strength can be adjusted by means of the valve angles u_{BS} and u_{SB}. The ratio of the volume that is used for the coupling of the systems and the outflow to TW is set by the valve angles u_{BW} and u_{SW}.

The nonlinear model of the process and the linearized model that is used for elaborating the proposed event-based control approaches are given in Appendices A.1 and A.2, respectively.

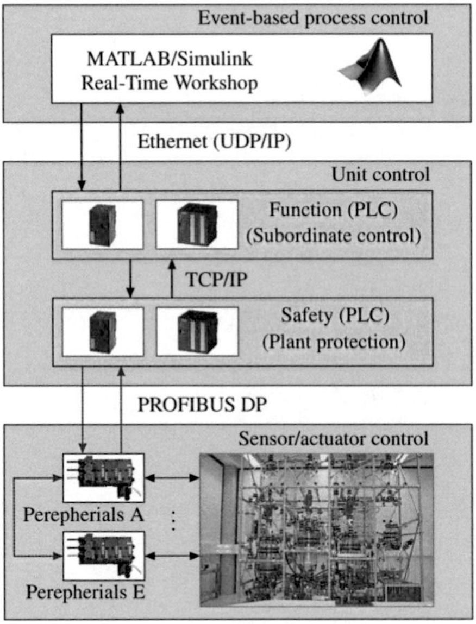

Figure 2.5: Automation concept for the pilot plant

Hardware description. Figure. 2.5 illustrates the automation concept for the pilot plant which is subdivided into three layers. On the top layer, the control algorithms for all event-based control approaches presented in this thesis are implemented in MATLAB/Simulink running on an ordinary personal computer (PC). MATLAB/Simulink is executed using Real-Time workshop with the sampling time $T_{\mathrm{s}} = 0.3\,\mathrm{s}$. As the sampling time T_{s} is by a factor of more than 150 smaller than the time constants of the process, the control can be considered to be continuous. The PC is connected over a 100 Mbit/s Ethernet network with the programmable logic controllers (PLCs) on which subordinate controllers and several routines for the plant protection are implemented. On the field level the actuator signals are applied and the sensor signals are sampled via the peripheral units that are connected over PROFIBUS DP with the PLCs.

3 State-feedback approach to event-based control

The state-feedback approach to event-based control published in [115] is based on the idea that the event-based control shall be made to mimic the behavior of a continuous state feedback with adjustable precision. This chapter summarizes this approach and presents its main properties. The behavior of the event-based state feedback applied to the thermofluid process is demonstrated through a simulation example.

3.1 Basic idea

The state-feedback approach to event-based control published in [115] is grounded on the consideration that feedback control, as opposed to feedforward control, is necessary in three situations:

- An unstable plant needs to be stabilized.
- The plant is inaccurately known so that the controller needs to react to model uncertainties.
- Unknown disturbances need to be attenuated.

In order to answer the question, at which time instants a control loop needs to be closed, the work [115] has been focused on the aspect of disturbance attenuation while the plant has been assumed to be stable and the plant model is known with negligible uncertainties. Hence, the only reason for a feedback communication is that the disturbance $d(t)$ has an intolerable effect on the system performance. Further publications that have followed up on [115] have extended the event-based state-feedback approach to deal with model uncertainties [100], delays and packet losses in the feedback communication [102] and nonlinear plant dynamics [2, 3, 7], though the analysis of the event-based control loop requires different methods in these cases.

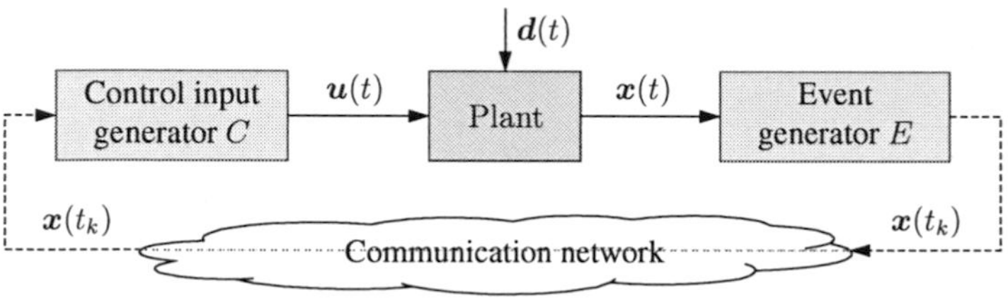

Figure 3.1: Event-based state-feedback loop

The structure of the event-based control loop investigated in [115] is illustrated in Fig. 3.1. The plant is represented by the model (2.1), assuming that it provides a central sensor unit and a central actuator unit. Regarding the plant the assumptions **A** 1.1–**A** 1.3 are made. The *event generator* E determines the event times t_k $(k = 0, 1, 2, \ldots)$, at which the current state $\boldsymbol{x}(t_k)$ is transmitted to the *control input generator* C, that uses the received information to update the generation of the control input $\boldsymbol{u}(t)$. The solid arrows represent continuous information links, whereas the dashed line symbolizes a communication that occurs at the event times only. The feedback communication via the network is assumed to be ideal as specified in **A** 1.6.

Remark 3.1 *The structure of the event-based state-feedback loop shown in Fig. 3.1 differs from the general structure of event-based control loops depicted in Fig. 1.1 in that the controller is incorporated in the control input generator C and the event generator E and, thus, does not appear explicitly as a separate block. By considering the controller to be included in these generators it is implicitly assumed that their processing power does not impose any limitations on the computations to be carried out.*

The basic idea of the event-based state-feedback approach is to design the event generator E and the control input generator C such that the event-based control loop imitates the disturbance behavior of a continuous state-feedback loop with arbitrarily adjustable approximation error. In the following this continuous control system is referred to as the *reference system* which is represented by the model

$$\Sigma_{\mathrm{r}}: \quad \dot{\boldsymbol{x}}_{\mathrm{r}}(t) = \underbrace{(\boldsymbol{A} - \boldsymbol{B}\boldsymbol{K})}_{=: \bar{\boldsymbol{A}}} \boldsymbol{x}_{\mathrm{r}}(t) + \boldsymbol{E}\boldsymbol{d}(t), \quad \boldsymbol{x}_{\mathrm{r}}(0) = \boldsymbol{x}_0 \tag{3.1}$$

with the state $\boldsymbol{x}_{\mathrm{r}} \in \mathbb{R}^n$. The state of the reference system is marked by the subscript r in order to distinguish it from the state of the event-based control system introduced later. The state-feedback gain $\boldsymbol{K}$ in (3.1) is assumed to be designed such that the system Σ_{r} has a desired disturbance behavior and, hence, the state $\boldsymbol{x}_{\mathrm{r}}(t)$ of the reference system (3.1) is bounded.

Proposition 3.1 *Given that the disturbance $\boldsymbol{d}(t)$ is bounded as in (2.2), the reference system*

(3.1) *is practically stable with respect to the set*

$$\mathcal{A}_\mathrm{r} := \{\boldsymbol{x}_\mathrm{r} \in \mathbb{R}^n \mid \|\boldsymbol{x}_\mathrm{r}\| \leq b_\mathrm{r}\} \tag{3.2a}$$

with the bound

$$b_\mathrm{r} = \|\bar{\boldsymbol{d}}\| \cdot \int_0^\infty \left\| \mathrm{e}^{\bar{\boldsymbol{A}}\tau} \boldsymbol{E} \right\| \mathrm{d}\tau. \tag{3.2b}$$

Proof. Equation (3.1) yields

$$\boldsymbol{x}_\mathrm{r}(t) = \mathrm{e}^{\bar{\boldsymbol{A}}t}\boldsymbol{x}_0 + \int_0^t \mathrm{e}^{\bar{\boldsymbol{A}}(t-\tau)}\boldsymbol{E}\boldsymbol{d}(\tau)\mathrm{d}\tau.$$

For an arbitrary large but bounded initial state $\boldsymbol{x}_0$ and the maximum disturbance magnitude $\bar{\boldsymbol{d}}$ the norm of the state $\|\boldsymbol{x}_\mathrm{r}(t)\|$ is bounded by

$$\limsup_{t\to\infty} \|\boldsymbol{x}_\mathrm{r}(t)\| \leq \|\bar{\boldsymbol{d}}\| \cdot \int_0^\infty \left\| \mathrm{e}^{\bar{\boldsymbol{A}}\tau} \boldsymbol{E} \right\| \mathrm{d}\tau =: b_\mathrm{r}$$

for $t \to \infty$. □

3.2 Components of the event-based controller

This section describes the method for the design of the event-based controller proposed in [115]. The event-based controller consists of the following components:

- The event generator E that determines what information is transmitted via the feedback link at which time instants t_k.

- The control input generator C which generates the control input $\boldsymbol{u}(t)$ in a feedforward manner in between consecutive event times t_{k-1} and t_k and it updates the determination of the input signal $\boldsymbol{u}(t)$ using the information received at the event times t_k.

The communication network is also considered to be part of the event-based controller. However, the network is assumed to be ideal as stated in **A** 1.6 and, thus, the transmission of data is not particularly considered.

3.2.1 Control input generator C

The control input generator C is described by

$$C : \begin{cases} \Sigma_s : & \dot{\boldsymbol{x}}_s(t) = \bar{\boldsymbol{A}}\boldsymbol{x}_s(t) + \boldsymbol{E}\hat{\boldsymbol{d}}_k, \quad \boldsymbol{x}_s(t_k^+) = \boldsymbol{x}(t_k) \\ & \boldsymbol{u}(t) = -\boldsymbol{K}\boldsymbol{x}_s(t). \end{cases} \tag{3.3}$$

It includes a model Σ_s of the reference system (3.1) with the model state $\boldsymbol{x}_s \in \mathbb{R}^n$ which is reset at the event times t_k to the current plant state $\boldsymbol{x}(t_k)$. $\hat{\boldsymbol{d}}_k$ denotes an estimate of the disturbance $\boldsymbol{d}(t)$ which is determined at the event time t_k. A method for the disturbance estimate is presented in Sec. 3.2.3. The control input $\boldsymbol{u}(t)$ is generated using the model state $\boldsymbol{x}_s(t)$ and is applied to the plant.

3.2.2 Event generator E

The event generator E also includes a model Σ_s of the reference system (3.1) and it determines the event times t_k as follows

$$E : \begin{cases} \Sigma_s : & \dot{\boldsymbol{x}}_s(t) = \bar{\boldsymbol{A}}\boldsymbol{x}_s(t) + \boldsymbol{E}\hat{\boldsymbol{d}}_k, \quad \boldsymbol{x}_s(t_k^+) = \boldsymbol{x}(t_k) \\ & t_0 = 0, \\ & t_{k+1} := \inf\left\{ t > t_k \mid \|\boldsymbol{x}(t) - \boldsymbol{x}_s(t)\| = \bar{e} \right\}. \end{cases} \tag{3.4}$$

Note that the state $\boldsymbol{x}_s(t)$ of the model Σ_s that is used by both the control input generator C and the event generator E is a prediction of the plant state $\boldsymbol{x}(t)$. Given that assumption **A** 1.1 holds true and that $\boldsymbol{x}_s(t_k^+) = \boldsymbol{x}(t_k)$, a deviation between $\boldsymbol{x}(t)$ and $\boldsymbol{x}_s(t)$ for $t > t_k$ is only caused by a difference between the actual disturbance $\boldsymbol{d}(t)$ and its estimate $\hat{\boldsymbol{d}}_k$. The event generator continuously monitors the difference state

$$\boldsymbol{x}_\Delta(t) := \boldsymbol{x}(t) - \boldsymbol{x}_s(t) \tag{3.5}$$

and triggers an event, whenever the norm of this difference attains some event threshold $\bar{e} \in \mathbb{R}_+$. The initial event at time $t_0 = 0$ is generated regardless of the triggering condition. At the event time t_k ($k = 0, 1, 2, \ldots$), the event generator E transmits the current plant state $\boldsymbol{x}(t_k)$ and a new disturbance estimate $\hat{\boldsymbol{d}}_k$ over the communication network to the control input generator C. The state information $\boldsymbol{x}(t_k)$ is used in both components at the event time t_k in order to reset the model state $\boldsymbol{x}_s(t)$, as indicated in (3.3), (3.4).

3.2.3 Disturbance estimation

The state-feedback approach to event-based control [115] works with any disturbance estimation method that yields bounded estimates $\hat{\boldsymbol{d}}_k$, including the trivial estimation $\hat{\boldsymbol{d}}_k \equiv 0$ for all $k = \mathbb{N}_0$. This section presents an estimation method which is based on the assumption that the disturbance $\boldsymbol{d}(t)$ is constant in the time interval $t \in [t_{k-1}, t_k)$

$$\boldsymbol{d}(t) = \boldsymbol{d}_{\mathrm{c}}, \quad \text{for } t \in [t_{k-1}, t_k)$$

where $\boldsymbol{d}_{\mathrm{c}} \in \mathcal{D}$ denotes the disturbance magnitude. The idea of this estimation method is to determine the disturbance magnitude $\boldsymbol{d}_{\mathrm{c}}$ at the event time t_k and to use this value as the disturbance estimation $\hat{\boldsymbol{d}}_k$ for the time $t \geq t_k$ until the next event occurs.

The disturbance estimation method is given by the following recursion

$$\hat{\boldsymbol{d}}_0 = 0 \tag{3.6a}$$

$$\hat{\boldsymbol{d}}_k = \hat{\boldsymbol{d}}_{k-1} + \left(\boldsymbol{A}^{-1}\left(\mathrm{e}^{\boldsymbol{A}(t_k - t_{k-1})} - \boldsymbol{I}_n\right)\boldsymbol{E}\right)^{+} \boldsymbol{x}_{\Delta}(t_k) \tag{3.6b}$$

where the initial estimation $\hat{\boldsymbol{d}}_0$ is chosen to be zero if no information about the disturbance is available. The pseudoinverse in (3.6) exists if $p \leq n$ holds, i.e., the dimension p of the disturbance vector $\boldsymbol{d}(t)$ does not exceed the dimension n of the state vector $\boldsymbol{x}(t)$.

Note that the estimate $\hat{\boldsymbol{d}}_k$ that is determined according to this estimation method is a weighted average of the disturbance $\boldsymbol{d}(t)$ in the preceding time interval $t \in [t_{k-1}, t_k)$. This property of the estimation method is illustrated in Fig. 3.2 for the example of a scalar disturbance $d(t)$, where the weighted average of the signal $d(t)$ in the respective time intervals is shown as a dotted line. Note that the disturbance $d(t)$ is estimated accurately by the method (3.6) if $d(t)$ remains constant for two consecutive time intervals [115, 98]. This situation is shown in Fig. 3.2 for $t \geq t_3$ where the disturbance $d(t)$ is estimated correctly at the event time t_4.

The disturbance estimation method is implemented in the event generator E as well as in the control input generator C. The event generator E transmits at the event time t_k the current plant state $\boldsymbol{x}(t_k)$ and this information is then used in the control input generator C to reset the

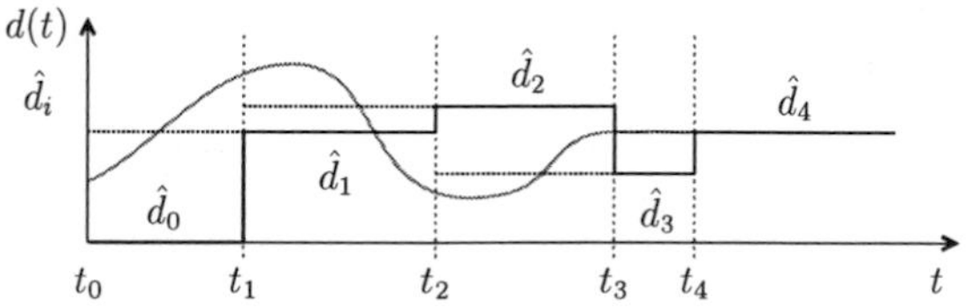

Figure 3.2: Disturbance $d(t)$ and disturbance estimates $\hat{d}_i$

model state $\boldsymbol{x}_\mathrm{s}$ and to determine the disturbance estimation. An alternative realization of the disturbance estimation is to implement the estimation method (3.6) in the event generator E only. In that case, E transmits the new disturbance estimation $\hat{\boldsymbol{d}}_k$ together with the current plant state $\boldsymbol{x}(t_k)$ at the event time t_k to the control input generator C.

3.3 Main properties of the event-based state feedback

3.3.1 Deviation between the behavior of the reference system and the event-based control loop

The basic idea of the event-based state-feedback approach [115] is to approximate the disturbance behavior of the continuous state-feedback system (3.1) by the event-based state feedback (2.1), (3.3), (3.4) with arbitrary precision. The next theorem shows that the approximation error

$$\boldsymbol{e}(t) := \boldsymbol{x}(t) - \boldsymbol{x}_\mathrm{r}(t)$$

is bounded and, moreover, can be adjusted by appropriately chosing the event threshold $\bar{e}$.

Theorem 3.1 (Theorem 1 in [115]) *The approximation error $\boldsymbol{e}(t)$ between the behavior of the event-based state-feedback loop (2.1), (3.3), (3.4) and the reference system (3.1) is bounded from above by*

$$\|\boldsymbol{e}(t)\| \leq \bar{e} \cdot \int_0^\infty \left\| \mathrm{e}^{\bar{\boldsymbol{A}}t} \boldsymbol{B}\boldsymbol{K} \right\| \mathrm{d}t =: e_{\max} \tag{3.7}$$

for all $t \geq 0$.

The theorem shows that the deviation between the disturbance rejection behavior of the event-based control system (2.1), (3.3), (3.4) and the reference system (3.1) depends linearly upon the event threshold $\bar{e}$. Hence, the behavior of the reference system is better approximated by the event-based state-feedback loop if the threshold $\bar{e}$ is reduced and vice versa. Theorem 3.1 can also be regarded as a proof of stability for the event-based state-feedback loop, since the stability of the reference system (3.1) together with the bound (3.7) implies the boundedness of the state $\boldsymbol{x}(t)$ of the event-based control loop.

The event-based state-feedback loop (2.1), (3.3), (3.4) is practically stable with respect to the set

$$\mathcal{A} := \{\boldsymbol{x} \in \mathbb{R}^n \mid \|\boldsymbol{x}\| \leq b\}$$

with the bound

$$b = \bar{e} \cdot \int_0^{\infty} \left\| \mathrm{e}^{\bar{\boldsymbol{A}}\tau} \boldsymbol{B}\boldsymbol{K} \right\| \mathrm{d}\tau + \|\bar{\boldsymbol{d}}\| \cdot \int_0^{\infty} \left\| \mathrm{e}^{\bar{\boldsymbol{A}}\tau} \boldsymbol{E} \right\| \mathrm{d}\tau, \tag{3.8}$$

(the derivation of the bound b can be found in [98]). A comparison of (3.8) with the bound b_{r} for the reference system given in (3.2) implies the relation

$$b = b_{\mathrm{r}} + e_{\max}.$$

Hence, the bound b_{r} on the set $\mathcal{A}_{\mathrm{r}}$ for the reference system is extended by the bound (3.7) if event-based state feedback is applied instead of continuous state feedback.

3.3.2 Adaption of the communication effort to the system behavior

This section shows that the event-based state feedback adapts the communication effort over the network to the system behavior. This can be seen by considering the difference state (3.5). The model Σ_{s} defined in (3.3) and the plant model (2.1) yield

$$\dot{\boldsymbol{x}}_{\Delta}(t) = \boldsymbol{A}\boldsymbol{x}_{\Delta}(t) + \boldsymbol{E}\boldsymbol{d}_{\Delta}(t), \quad \boldsymbol{x}_{\Delta}(t_k^+) = \boldsymbol{0},$$

where $\boldsymbol{d}_{\Delta}(t) = \boldsymbol{d}(t) - \hat{\boldsymbol{d}}_k$ denotes the deviation between the actual disturbance $\boldsymbol{d}(t)$ and the estimate $\hat{\boldsymbol{d}}_k$ applied in the model Σ_{s} for $t \geq t_k$. For the time $t \geq t_k$ the previous equation results in

$$\boldsymbol{x}_{\Delta}(t) = \int_{t_k}^{t} \mathrm{e}^{\boldsymbol{A}(t-\tau)} \boldsymbol{E}\boldsymbol{d}_{\Delta}(\tau)\mathrm{d}\tau.$$

Recall that, according to the event condition given in (3.4), an event is triggered whenever the relation

$$\|\boldsymbol{x}_{\Delta}(t)\| = \left\| \int_{t_k}^{t} \mathrm{e}^{\boldsymbol{A}(t-\tau)} \boldsymbol{E}\boldsymbol{d}_{\Delta}(\tau)\mathrm{d}\tau \right\| = \bar{e}$$

is satisfied, which shows that events are only triggered if the signal $\boldsymbol{d}_{\Delta}(t)$ is sufficiently large, i.e., the actual disturbance $\boldsymbol{d}(t)$ considerably deviates from the estimate $\hat{\boldsymbol{d}}_k$. Hence, no feedback communication is necessary for $t \geq t_k$ if the signal $\boldsymbol{d}_{\Delta}(t)$ is small enough, [5, 115]. The next lemma makes this statement more precise considering the case that no events, except for the initial event at time $t_0 = 0$, are triggered at all if the disturbance magnitude $\bar{\boldsymbol{d}}$ is sufficiently small.

Lemma 3.1 (Lemma 3 in [115]) *The event generator E does not trigger an event for $t > 0$ if the disturbance magnitude $\bar{\boldsymbol{d}}$ satisfies the relation*

$$\|\bar{\boldsymbol{d}}\| < \bar{e} \cdot \left(\int_0^\infty \left\| \mathrm{e}^{\boldsymbol{A}\tau} \boldsymbol{E} \right\| \mathrm{d}\tau \right)^{-1}. \tag{3.9}$$

Relation (3.9) gives a quantitative bound on the disturbance magnitude $\bar{\boldsymbol{d}}$ for which the feedback loop does not need to be closed in order to guarantee a desired control performance. In other words, the disturbance magnitude $\bar{\boldsymbol{d}}$ is so small that the disturbance $\boldsymbol{d}(t)$ remains undetected by the event generator E.

3.3.3 Minimum inter-event time

The minimum time that elapses in between two consecutive events is referred to as the *minimum inter-event time* $T_{\min}$. The event-based state-feedback approach [115] ensures that the minimum inter-event time is bounded from below according to $T_{\min} \geq \bar{T}$.

Theorem 3.2 (Theorem 2 in [115]) *For any bounded disturbance, the minimum inter-event time $T_{\min}$ of the event-based state-feedback loop (2.1), (3.3), (3.4) is bounded from below by*

$$\bar{T} = \arg\min_t \left\{ \int_0^t \left\| \mathrm{e}^{\boldsymbol{A}\tau} \boldsymbol{E} \right\| \mathrm{d}\tau = \frac{\bar{e}}{\bar{d}_\Delta} \right\} \tag{3.10}$$

where $\bar{d}_\Delta \geq \left\| \boldsymbol{d}(t) - \hat{\boldsymbol{d}}_k \right\|$ denotes the maximum deviation between the disturbance $\boldsymbol{d}(t)$ and the estimates $\hat{\boldsymbol{d}}_k$ for all $t \geq 0$ and all $k \in \mathbb{N}_0$.

A direct inference of Theorem 3.2 is that no Zeno behavior can occur in the event-based state-feedback loop, that is the number of events does not accumulate to infinity in finite time [114]. This result also shows that the minimum inter-event time gets smaller if the event threshold $\bar{e}$ is reduced and vice versa. This conclusion is in accordance with the expectation that a more precise approximation of the reference systems behavior by the event-based state feedback comes at the cost of a more frequent communication.

3.4 Example: Event-based state-feedback control of the thermofluid process

This section demonstrates the behavior of the event-based state-feedback approach using the example of the thermofluid process, introduced in Sec. 2.4. For this example the process is considered to have a centralized sensor unit providing the measurements of the levels $l_{\mathrm{B}}(t)$ and

$l_{\mathrm{S}}(t)$ as well as the temperatures $\vartheta_{\mathrm{B}}(t)$ and $\vartheta_{\mathrm{S}}(t)$ and a centralized actuator unit that applies the overall control input $\boldsymbol{u}(t)$.

The aim of the event-based state-feedback control is to approximate the behavior of a continuous state feedback with the feedback gain

$$\boldsymbol{K} = \begin{pmatrix} 10.72 & 0.13 & -0.21 & 0.08 \\ 0.15 & -0.14 & -0.02 & -0.14 \\ -0.14 & 0.31 & 11.94 & 0.12 \\ -0.63 & 0.17 & 1.21 & 0.42 \end{pmatrix}. \tag{3.11}$$

The state $\boldsymbol{x}_{\mathrm{s}}(t)$ of the model Σ_{s} used in the event generator E as well as in the control input generator C is denoted by

$$\boldsymbol{x}_{\mathrm{s}}^{\top}(t) = \begin{pmatrix} l_{\mathrm{sB}}(t) & \vartheta_{\mathrm{sB}}(t) & l_{\mathrm{sS}}(t) & \vartheta_{\mathrm{sS}}(t) \end{pmatrix}.$$

The event threshold is set to $\bar{e} = 2$, taking into account that the levels $l_{\mathrm{B}}(t)$ and $l_{\mathrm{S}}(t)$ are converted into cm before evaluating the event condition. Hence, an event is triggered whenever the equality

$$\left\| \begin{pmatrix} 100 \cdot (l_{\mathrm{B}}(t) - l_{\mathrm{sB}}(t)) \\ \vartheta_{\mathrm{B}}(t) - \vartheta_{\mathrm{sB}}(t) \\ 100 \cdot (l_{\mathrm{S}}(t) - l_{\mathrm{sS}}(t)) \\ \vartheta_{\mathrm{S}}(t) - \vartheta_{\mathrm{sS}}(t) \end{pmatrix} \right\|_{\infty} = \bar{e} = 2$$

holds true. Note that the uniform norm is applied here, which means that the model state $\boldsymbol{x}_{\mathrm{s}}(t)$ is reset to the current plant state $\boldsymbol{x}(t)$ whenever the level or the temperature in reactor TB or in reactor TS deviates by $2\,\mathrm{cm}$ or by $2\,\mathrm{K}$ from the model counterpart. With the chosen state-feedback gain $\boldsymbol{K}$ and event threshold $\bar{e}$ the maximum approximation error

$$e_{\mathrm{max}} = 12.01$$

results from Eq. (3.7). Moreover, with $\bar{d}_{\Delta} = 0.5$,

$$\bar{T} = 26\,\mathrm{s}$$

is obtained from Eq. (3.10) as the lower bound on the minimum inter-event time T_{min}.

Figure 3.3 depicts the disturbance behavior of the event-based state-feedback system, where the following scenario is investigated: The reactor TB is perturbed in the time interval $t \in$

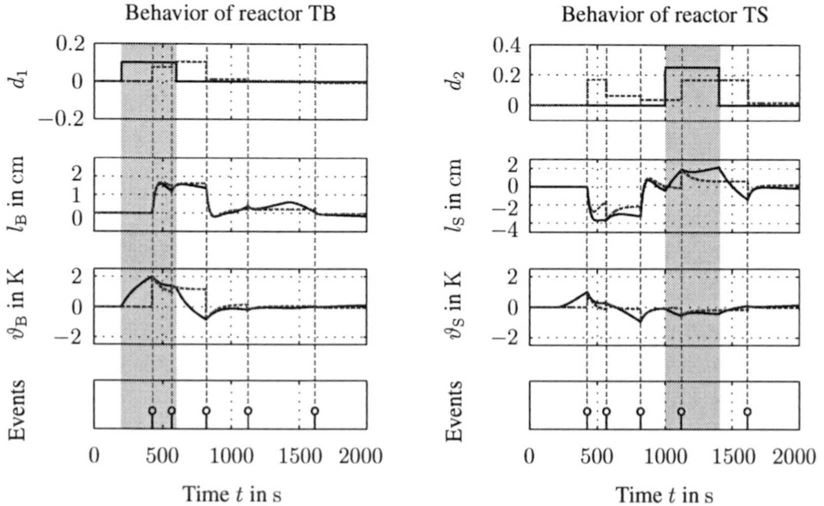

Figure 3.3: Disturbance behavior of the event-based state-feedback approach. The behavior of reactor TB and reactor TS is shown on the left-hand side or right-hand side, respectively. The first row shows the disturbances (solid lines) and its estimation (dashed lines). The trajectories of the level and temperature are given in the second and third row (solid line: plant state $\boldsymbol{x}(t)$, dashed line: model state $\boldsymbol{x}_s(t)$). The event time instants are represented by stems in the bottom figure.

$[200, 600]$ s by the activation of the heating rods with $d_1(t) = d_H(t) = 0.1$ which has an immediate effect on the temperature $\vartheta_B(t)$ of the liquid in TB. In the time interval $t \in [1000, 1400]$ s an additional inflow $d_2(t) = d_F(t) = 0.25$ affects both the level $l_S(t)$ and the temperature $\vartheta_S(t)$ in reactor TS. The time intervals where a disturbance is active are highlighted in gray in Fig. 3.3. Note that the disturbances are propagated via the coupling through the whole system, since both reactors are physically interconnected.

The behavior of the event-based control system shown in Fig. 3.3 is representative for the adaption of the feedback communication to the system behavior. Events are triggered only if a disturbance is active or if the disturbance estimate deviates significantly from the actual disturbance. The minimum time that elapses in between two consecutive events observed in this example is

$$T_{\min} = 139\,\text{s}$$

and, thus, is much larger than the lower bound $\bar{T} = 26$ s.

Figure 3.4 illustrates the trajectories of the difference state $\boldsymbol{x}_\Delta(t) = \boldsymbol{x}(t) - \boldsymbol{x}_s(t)$. The

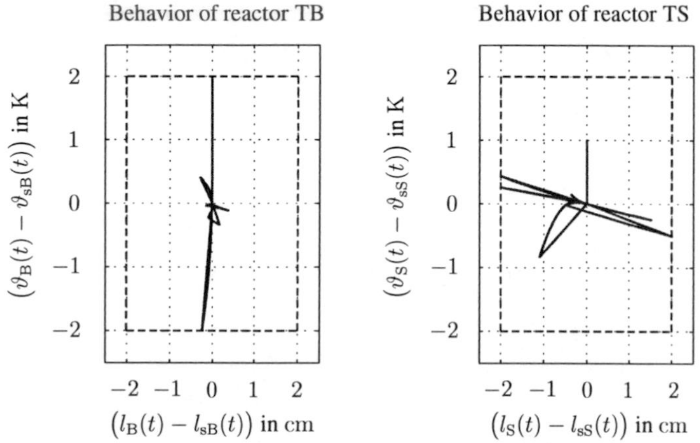

Figure 3.4: Trajectories of the difference states $\boldsymbol{x}_{\Delta 1}(t)$ and $\boldsymbol{x}_{\Delta 2}(t)$. The dashed lines represents the event threshold $\bar{e} = 2$.

difference state $\boldsymbol{x}_{\Delta}(t)$ is reset to zero, whenever one of the trajectories in both figures attains a dashed line, which symbolizes the event threshold $\bar{e} = 2$. From this figure it can be inferred that events are only triggered due to the deviation of the temperature $\vartheta_B(t)$ or the level $l_S(t)$ from the corresponding predictions $\vartheta_{sB}(t)$ or $l_{sS}(t)$, respectively.

Finally, Fig. 3.5 depicts the deviation between the behavior of the event-based state-feedback loop and the reference system with continuous state feedback. The maximum approximation error is

$$\max_{t \geq 0} \|\boldsymbol{e}(t)\|_{\infty} = 1.03$$

which is by a factor of more than 11 smaller than the derived bound $e_{\max} = 12.01$. This investigation indicates that the determination of the bound on the maximum approximation error according to (3.7) might yield conservative results.

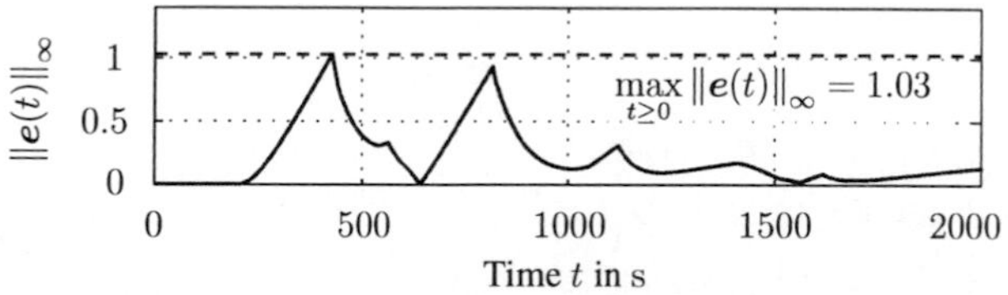

Figure 3.5: Deviation $\boldsymbol{e}(t)$ between the behavior of the event-based state-feedback loop and the reference system.

4 Event-based control of interconnected systems

This chapter presents a concept for the extension of the event-based state-feedback approach from systems with centralized sensor and actuator units to physically interconnected systems with multiple sensor and actuator units. A general structure for an event-based controller for interconnected systems is introduced that consists of decentralized local control units which communicate over a communication network. The question is investigated what conditions an event-based controller must fulfill in order to guarantee that the event-based control loop approximates the behavior of a continuous reference system with adjustable accuracy.

4.1 Structures of event-based controllers for interconnected systems

The event-based state-feedback approach [115] is applicable to multivariable systems, but requires simultaneous measurements of all sensors and simultaneous updates of the control inputs by all actuators. Hence, the applicability of the event-based state-feedback approach is limited to systems that satisfy this prerequisite. Most technical systems, however, consist of several subsystems Σ_i that are physically interconnected, e.g., chemical process plants [1, 143] or power system [22, 72, 113]. These kind of systems are generally large-scale systems which do not provide a central sensor unit and a central actuator unit. Instead, each subsystem Σ_i has a local sensor unit that measures the subsystem state $\boldsymbol{x}_i(t)$ and a local actuator unit that applies the control input $\boldsymbol{u}_i(t)$ (Fig. 2.1(b)).

The event-based control approaches that are proposed in this thesis extend the event-based state-feedback approach [115] to physically interconnected systems (2.3), (2.4),

$$\Sigma_i : \begin{cases} \dot{\boldsymbol{x}}_i(t) = \boldsymbol{A}_i\boldsymbol{x}_i(t) + \boldsymbol{B}_i\boldsymbol{u}_i(t) + \boldsymbol{E}_i\boldsymbol{d}_i(t) + \boldsymbol{E}_{si}\boldsymbol{s}_i(t), \quad \boldsymbol{x}_i(0) = \boldsymbol{x}_{0i} \\ \boldsymbol{z}_i(t) = \boldsymbol{C}_{zi}\boldsymbol{x}_i(t) \end{cases} \tag{4.1a}$$

$$\boldsymbol{s}_i(t) = \sum_{j=1}^{N} \boldsymbol{L}_{ij}\boldsymbol{z}_j(t), \tag{4.1b}$$

retaining the property that the behavior of the reference system (3.1)

$$\Sigma_r : \quad \dot{\boldsymbol{x}}_r(t) = (\boldsymbol{A} - \boldsymbol{B}\boldsymbol{K})\,\boldsymbol{x}_r(t) + \boldsymbol{E}\boldsymbol{d}(t), \quad \boldsymbol{x}_r(0) = \boldsymbol{x}_0 \tag{4.2}$$

is approximated by the event-based control loop with adjustable accuracy. The aim of this section is to identify structures of event-based state-feedback controllers for interconnected systems which allow for an adjustable deviation between the behavior of the reference system and the event-based control loop.

The starting point of this investigation is the general structure of the event-based control loop for interconnected systems shown in Fig. 4.1. The overall plant is controlled by means of a networked controller F consisting of local control units F_i that are implemented in a decentralized manner and which can communicate over a network. Note that the networked controller F represents an abstraction for the structures of the event-based controllers for interconnected systems that are investigated in the next chapters of this thesis. Hence, the local control units F_i are considered to be event-based controllers which consist of an event generator E_i and a control input generator C_i. Based on this structure of the control loop the following question is investigated:

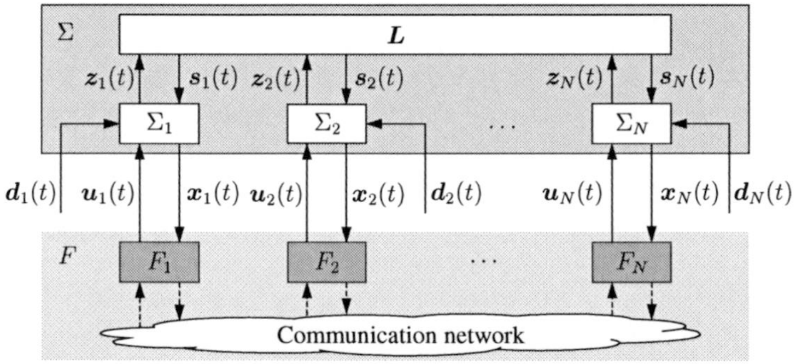

Figure 4.1: General structure of the event-based control system

> What conditions must an event-based controller F for the interconnected systems (4.1a), (4.1b) fulfill in order to ensure that the deviation between the behavior of the event-based state-feedback loop and the continuous reference system (4.2) is adjustable?

In order to answer this question assume that the local control unit F_i is represented by the model

$$F_i : \begin{cases} \boldsymbol{x}_{\mathrm{s}}^i(t) = \boldsymbol{g}_i(\cdot), \quad \boldsymbol{x}_{\mathrm{s}}^i(t_k^+) = \boldsymbol{f}_i(\boldsymbol{x}(t_k)), & \text{for } t \geq t_k \\ \boldsymbol{u}_i(t) = -\boldsymbol{K}_i \boldsymbol{x}_{\mathrm{s}}^i(t). \end{cases} \tag{4.3}$$

In (4.3), $\boldsymbol{x}_{\mathrm{s}}^i \in \mathbb{R}^n$ denotes a model state that can be interpreted as a prediction of the plant state $\boldsymbol{x}(t)$. In order to emphasize that this prediction is determined by the controller F_i the state is marked with a superscript i. For the subsequent investigation it is irrelevant how the state $\boldsymbol{x}_{\mathrm{s}}^i(t)$ is determined by F_i and, hence, the function $\boldsymbol{g}_i$ is left unspecified. $\boldsymbol{f}_i(\cdot)$ denotes the function that describes how the prediction $\boldsymbol{x}_{\mathrm{s}}^i(t)$ is reinitialized at the event times t_k. F_i generates the control input $\boldsymbol{u}_i(t)$ using the state-feedback gain $\boldsymbol{K}_i \in \mathbb{R}^{m_i \times n}$ where

$$\boldsymbol{K} = \begin{pmatrix} \boldsymbol{K}_1 \\ \vdots \\ \boldsymbol{K}_N \end{pmatrix} \tag{4.4}$$

is the state-feedback gain applied in the reference system (4.2). The event-based state-feedback loop is represented by the interconnected subsystems (4.1a), (4.1b) together with the event-based controllers (4.3). Note that the triggering mechanism of the controller F_i is deliberately left open in order to derive the condition on the state-feedback gain $\boldsymbol{K}$ and the communication structure. The following theorem provides an answer to the initial question [10].

Theorem 4.1 *The deviation $\mathbf{e}(t) = \boldsymbol{x}(t) - \boldsymbol{x}_{\mathrm{r}}(t)$ between the behavior of the event-based control system* (4.1a), (4.1b), (4.3) *and the reference system* (4.2) *is bounded, if the relation*

$$\sup_{t \geq 0} \left\| \boldsymbol{K}_i \left(\boldsymbol{x}(t) - \boldsymbol{x}_{\mathrm{s}}^i(t) \right) \right\| < \infty \tag{4.5}$$

holds for all $i \in \mathcal{N}$, $\boldsymbol{d} \in \mathcal{D}$ and for all $t \geq 0$.

Proof. Consider the local control units F_i which jointly generate the overall control input

$$\boldsymbol{u}(t) = \begin{pmatrix} \boldsymbol{u}_1(t) \\ \vdots \\ \boldsymbol{u}_N(t) \end{pmatrix} = - \begin{pmatrix} \boldsymbol{K}_1 \boldsymbol{x}_{\mathrm{s}}^1(t) \\ \vdots \\ \boldsymbol{K}_N \boldsymbol{x}_{\mathrm{s}}^N(t) \end{pmatrix}.$$

The overall plant (4.1a), (4.1b) subject to this control input is represented by the state-space model

$$\dot{\boldsymbol{x}}(t) = \boldsymbol{A}\boldsymbol{x}(t) - \boldsymbol{B}\begin{pmatrix} \boldsymbol{K}_1\boldsymbol{x}_{\mathrm{s}}^1(t) \\ \vdots \\ \boldsymbol{K}_N\boldsymbol{x}_{\mathrm{s}}^N(t) \end{pmatrix} + \boldsymbol{E}\boldsymbol{d}(t), \quad \boldsymbol{x}(0) = \boldsymbol{x}_0,$$

where the matrices $\boldsymbol{A}, \boldsymbol{B}$ and $\boldsymbol{E}$ are determined from the parameters of the subsystems according to (2.6). With

$$\boldsymbol{K}\boldsymbol{x}(t) = \begin{pmatrix} \boldsymbol{K}_1\boldsymbol{x}(t) \\ \vdots \\ \boldsymbol{K}_N\boldsymbol{x}(t) \end{pmatrix}$$

the last equation can be restated as

$$\dot{\boldsymbol{x}}(t) = \bar{\boldsymbol{A}}\boldsymbol{x}(t) + \boldsymbol{B}\begin{pmatrix} \boldsymbol{K}_1\big(\boldsymbol{x}(t) - \boldsymbol{x}_{\mathrm{s}}^1(t)\big) \\ \vdots \\ \boldsymbol{K}_N\big(\boldsymbol{x}(t) - \boldsymbol{x}_{\mathrm{s}}^N(t)\big) \end{pmatrix} + \boldsymbol{E}\boldsymbol{d}(t), \quad \boldsymbol{x}(0) = \boldsymbol{x}_0. \tag{4.6}$$

Now consider the approximation error $\boldsymbol{e}(t) = \boldsymbol{x}(t) - \boldsymbol{x}_{\mathrm{r}}(t)$ that describes the deviation between the state $\boldsymbol{x}(t)$ of the event-based state-feedback system (4.6) and the reference system (4.2) which yields

$$\dot{\boldsymbol{e}}(t) = \bar{\boldsymbol{A}}\boldsymbol{e}(t) + \boldsymbol{B}\begin{pmatrix} \boldsymbol{K}_1\big(\boldsymbol{x}(t) - \boldsymbol{x}_{\mathrm{s}}^1(t)\big) \\ \vdots \\ \boldsymbol{K}_N\big(\boldsymbol{x}(t) - \boldsymbol{x}_{\mathrm{s}}^N(t)\big) \end{pmatrix}, \quad \boldsymbol{e}(0) = \boldsymbol{0}. \tag{4.7}$$

Since the matrix $\bar{\boldsymbol{A}}$ is Hurwitz by design, the deviation error

$$\begin{aligned} \|\boldsymbol{e}(t)\| &= \left\| \int_0^t \mathrm{e}^{\bar{\boldsymbol{A}}(t-\tau)}\boldsymbol{B}\begin{pmatrix} \boldsymbol{K}_1\big(\boldsymbol{x}(\tau) - \boldsymbol{x}_{\mathrm{s}}^1(\tau)\big) \\ \vdots \\ \boldsymbol{K}_N\big(\boldsymbol{x}(\tau) - \boldsymbol{x}_{\mathrm{s}}^N(\tau)\big) \end{pmatrix} \mathrm{d}\tau \right\| \\ &\leq \int_0^t \left\| \mathrm{e}^{\bar{\boldsymbol{A}}(t-\tau)}\boldsymbol{B} \right\| \left\| \begin{pmatrix} \boldsymbol{K}_1\big(\boldsymbol{x}(\tau) - \boldsymbol{x}_{\mathrm{s}}^1(\tau)\big) \\ \vdots \\ \boldsymbol{K}_N\big(\boldsymbol{x}(\tau) - \boldsymbol{x}_{\mathrm{s}}^N(\tau)\big) \end{pmatrix} \right\| \mathrm{d}\tau \end{aligned}$$

remains bounded if the signals

$$\boldsymbol{K}_i\big(\boldsymbol{x}(t) - \boldsymbol{x}_{\mathrm{s}}^i(t)\big), \quad \forall\, i = 1, \ldots, N$$

remain bounded for all $t \geq 0$, which completes the proof. □

Note that the relation (4.5) in Theorem 4.1 is a sufficient condition for the boundedness of the deviation between the behavior of the reference system (4.2) and the event-based control system (4.1a), (4.1b), (4.3). It claims that the deviation

$$\boldsymbol{u}_{\Delta i}(t) = \boldsymbol{K}_i\big(\boldsymbol{x}(t) - \boldsymbol{x}_{\mathrm{s}}^i(t)\big)$$

between the event-based state feedback $\boldsymbol{u}_i(t) = -\boldsymbol{K}_i\boldsymbol{x}_{\mathrm{s}}^i(t)$ and the control input, which would be obtained by using a continuous state feedback, is bounded. The condition (4.5) can be satisfied for a given state-feedback gain $\boldsymbol{K}$ by means of an appropriate triggering mechanism in the local control units F_i together with an appropriately chosen topology for the communication between the control units. This implies that the condition (4.5) imposes requirements on the structure of the event-based state-feedback loop for interconnected systems. The event-based control methods that are presented in the next chapters satisfy the condition (4.5) in various ways:

- Chapter 5 proposes a method for event-based state-feedback control of interconnected systems where the state-feedback gain $\boldsymbol{K}$ is assumed to be a dense matrix, as it is usually obtained by means of a centralized design that does not take any restrictions on the structure of the feedback gain into account. In order to realize this state-feedback law in the framework of the proposed decentralized event-based controller structure, each local control unit F_i includes a model of the overall reference system that is used to compute a prediction $\boldsymbol{x}_{\mathrm{s}}^i(t)$ of the overall plant state $\boldsymbol{x}(t)$. At the event times t_k, which are determined by the local control units based on locally available information only, the respective subsystem state is broadcasted to all other control units. In this way the prediction $\boldsymbol{x}_{\mathrm{s}}^i(t)$ is identical in all local control units F_i $(i \in \mathcal{N})$ and its deviation to the actual plant state is bounded due to the triggering mechanism.

- In Chapter 6 it is assumed that the state-feedback gain $\boldsymbol{K} = \boldsymbol{K}_{\mathrm{d}}$ to be implemented in an event-based fashion is a decentralized controller. Each local control unit F_i of the event-based controller F incorporates a model of the related subsystem to determine a prediction $\boldsymbol{x}_{\mathrm{s}i}(t)$ of the local subsystem state $\boldsymbol{x}_i(t)$ and generating the local control input $\boldsymbol{u}_i(t) = -\boldsymbol{K}_{\mathrm{d}i}\boldsymbol{x}_{\mathrm{s}i}(t)$. In this way the fulfillment of the condition (4.5) only depends upon the difference $\boldsymbol{x}_i(t) - \boldsymbol{x}_{si}(t)$ and the boundedness of this difference is guaranteed by means of a triggering conditions which requires locally available information only.

- Chapter 7 presents a method for distributed event-based state feedback, where the state-feedback gain $\boldsymbol{K}$ is assumed to be chosen such that the control input for subsystem Σ_i depends upon the local state as well as on the state of some neighboring subsystems. For the realization of this control law in an event-based manner, each control unit F_i determines a prediction of the neighboring states in order to generate the local control input. The main problem which the design of the triggering mechanism has to address is that the condition (4.5) cannot be checked locally in the control units F_i. In order to still guarantee the fulfillment of this condition, a new kind of event-based control is introduced where an event triggers either the transmission of local information to or the request of information from the neighboring subsystems.

4.2 Event triggering in interconnected systems

In the presented approaches to event-based control of physically interconnected systems events are triggered by the local control units F_i based on locally available information. Hence, the triggering of events occurs asynchronously, which requires a distinction of the of the events that are triggered by the respective controllers F_i (Fig. 4.2). In the following, $k_i \in \mathbb{N}_0$ denotes the counter of the events that are generated by the controller F_i, while $k \in \mathbb{N}_0$ still denotes the global event counter. Accordingly, t_{k_i} is the time at which the k_i-th event is triggered by F_i.

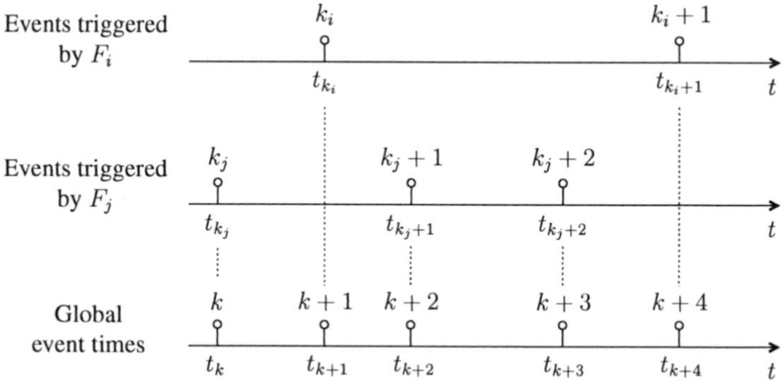

Figure 4.2: Asynchronuous event triggering

5 Event-based state-feedback control using broadcast communication

This chapter investigates methods for the event-based control of physically interconnected systems which use broadcast communication at the event times for the information exchange between the control units. A method for the distributed realization of the centralized event-based state-feedback approach [115] is proposed. It is shown that this distributed realization can be made to approximate the reference systems behavior with the same precision as the centralized event-based state feedback. The last part of this chapter presents an extension of the event-based state-feedback approach to systems where only a part of the overall state is measurable but still a desired level of performance should be accomplished.

5.1 Event-based control with broadcast communication

In networked control systems, the communication network introduces the flexibility of exchanging information between arbitrary components of the control system which is often not possible in conventional control loops with hard-wired connections between sensors, controller and actuators. Many bus-based communication systems, like CAN or PROFIBUS, even support a broadcast communication where the transmitted messages are received by all nodes in the network. The event-based control methods that are proposed in this chapter are designed for networked systems which feature a broadcast communication. In the context of event-based control it means that an event generator transmits at the event time the current state information to all other components of the networked controller, as illustrated in Fig. 5.1 for the example of the event generator E_2.

The possibility of broadcasting information brings up the question how the distribution of information among the entire network can be used to guarantee a desired performance of the

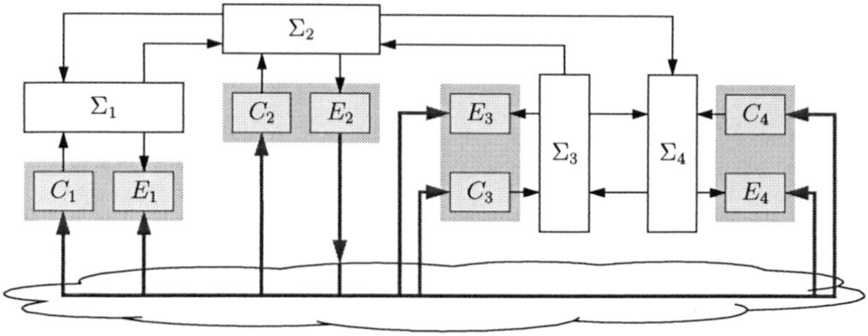

Figure 5.1: Broadcast communication scheme

overall control system. This issue is investigated in this chapter for two approaches to event-based control of interconnected systems:

- Section 5.2 proposes the distributed realization of the event-based state feedback which is a method to make the event-based state-feedback approach [115] applicable to systems that are composed of physically interconnected subsystems. Despite the decentralized controller structure, the distributed realization of the event-based state feedback is shown to approximate the behavior of the reference system (5.2) with adjustable accuracy. This control concept is tested on the thermofluid process through simulations and experiments, presented in Sec. 5.3.

- The basic concept of the distributed realization of the event-based state feedback is extended in Sec. 5.4 to the case where only a part of the overall state is measurable. Though the state is incompletely known, the overall event-based control system should approximate the behavior of a continuous state feedback with prescribed accuracy. The basic idea of the solution to this problem is to refine the information about the measurable states by choosing appropriate event thresholds in order to compensate for the loss of information about the unknown states. As the main result, an algorithm is presented that tests whether such event thresholds can be found and, if so, determines the thresholds using a linear programming approach.

5.2 Distributed realization of the event-based state-feedback approach

The main drawback of the event-based state-feedback approach [115] with respect to the applicability to large-scale systems is that the plant is required to have a central sensor unit and a central actuator unit. This section presents a method for the distributed realization of the event-based state feedback that is applicable to interconnected systems (2.3), (2.4)

$$\Sigma_i : \begin{cases} \dot{\boldsymbol{x}}_i(t) = \boldsymbol{A}_i \boldsymbol{x}_i(t) + \boldsymbol{B}_i \boldsymbol{u}_i(t) + \boldsymbol{E}_i \boldsymbol{d}_i(t) + \boldsymbol{E}_{\mathrm{s}i} \boldsymbol{s}_i(t), \quad \boldsymbol{x}_i(0) = \boldsymbol{x}_{0i} \\ \boldsymbol{z}_i(t) = \boldsymbol{C}_{\mathrm{z}i} \boldsymbol{x}_i(t) \end{cases} \tag{5.1a}$$

$$\boldsymbol{s}_i(t) = \sum_{j=1}^{N} \boldsymbol{L}_{ij} \boldsymbol{z}_j(t). \tag{5.1b}$$

Despite the decentralized controller structure, the proposed event-based control method preserves the property that the behavior of the reference system (3.1)

$$\Sigma_{\mathrm{r}} : \quad \dot{\boldsymbol{x}}_{\mathrm{r}}(t) = (\boldsymbol{A} - \boldsymbol{B}\boldsymbol{K})\, \boldsymbol{x}_{\mathrm{r}}(t) + \boldsymbol{E}\boldsymbol{d}(t), \quad \boldsymbol{x}_{\mathrm{r}}(0) = \boldsymbol{x}_0 \tag{5.2}$$

can be approximated by the event-based control system with arbitrary precision.

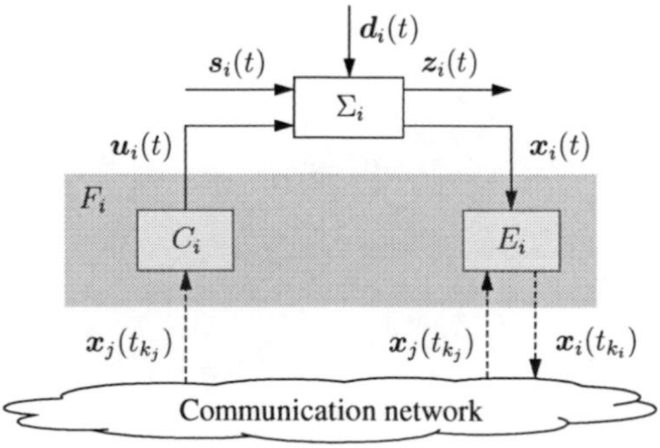

Figure 5.2: The local control unit F_i

5.2.1 Problem statement

The aim of the event-based controller that is presented in this section is to approximate the disturbance behavior of the reference system (5.2) with arbitrary precision. With respect to the reference system (5.2) no restrictions are imposed on the structure of the feedback gain $\boldsymbol{K}$. In the most general case the feedback gain $\boldsymbol{K}$ can be a dense matrix, as it might be obtained by means of a centralized controller design for the overall plant (2.1). This consideration implies that in the reference system (5.2) each component $\boldsymbol{u}_i(t)$ $(i = 1, \ldots, N)$ of the control input $\boldsymbol{u}(t)$ is a function of the overall plant state $\boldsymbol{x}(t)$.

The structure of the event-based control system investigated in this section is illustrated in Fig. 4.1 where for each $i \in \mathcal{N}$ the local control unit F_i consist of two components (Fig. 5.2): the control input generator C_i and the event generator E_i. The event generator E_i continuously measures the subsystem state $\boldsymbol{x}_i(t)$ and determines the event times t_{k_i} at which it transmits the current state $\boldsymbol{x}_i(t_{k_i})$ over the network to all other components of the controller F. For the determination of the event times t_{k_i}, the event generator E_i also uses the information $\boldsymbol{x}_j(t_{k_j})$ received from the other event generators E_j, $j \in \mathcal{N}, j \neq i$. The control input generator C_i uses the received information $\boldsymbol{x}_j(t_{k_j})$ with $j \in \mathcal{N}$ (including the case $j = i$) in order to determine the local control input $\boldsymbol{u}_i(t)$. Note that the control input generator C_i and the event generator E_i have no continuous connection. Hence, the sensor unit and the actuator unit of each subsystem Σ_i can be spatially distributed among the plant. The design of the control input generator C_i and the event generator E_i is explained in the next section.

The main problem that has to be dealt with by the event-based controller design is that a centralized state feedback shall be emulated by an event-based state feedback with a decentralized controller structure where, moreover, none of the control input generators has continuous access to any subsystem state.

5.2.2 Description of the components

The basic idea of the solution to the previously described design problem is to include in each event generator and control input generator the model

$$\Sigma_{\mathrm{s}}: \begin{cases} \dot{\boldsymbol{x}}_{\mathrm{s}}(t) = \bar{\boldsymbol{A}}\boldsymbol{x}_{\mathrm{s}}(t) + \boldsymbol{E}\hat{\boldsymbol{d}}_k \\ \boldsymbol{x}_{\mathrm{s}}(t_{k_i}^+) = \left(\boldsymbol{x}_{\mathrm{s}1}^\top(t_{k_i}) \ \ldots \ \boldsymbol{x}_{\mathrm{s},i-1}^\top(t_{k_i}) \ \boldsymbol{x}_i^\top(t_{k_i}) \ \boldsymbol{x}_{\mathrm{s},i+1}^\top(t_{k_i}) \ \ldots \ \boldsymbol{x}_{\mathrm{s}N}^\top(t_{k_i})\right)^\top \end{cases} \tag{5.3}$$

of the overall reference system (5.2) in order to determine the prediction $\boldsymbol{x}_{\mathrm{s}}(t)$ of the overall plant state $\boldsymbol{x}(t)$ where $\hat{\boldsymbol{d}}_k$ denotes an estimate of the disturbance $\boldsymbol{d}(t)$. At the event times t_{k_i} $(i \in \mathcal{N})$ only the i-th component $\boldsymbol{x}_{\mathrm{s}i}(t)$ of the model state $\boldsymbol{x}_{\mathrm{s}}(t)$ is reset and the disturbance estimation $\hat{\boldsymbol{d}}_k$ is updated at the same time using the same information in all components. Owing to the broadcast communication the models (5.3) are synchronized for all $t \geq 0$ in all generators. Consequently, a distinction of the model states using a superscript (as in the condition (4.5)) is omitted.

Each local control unit F_i $(i \in \mathcal{N})$ consists of a control input generator C_i and an event generator E_i which are subsequently described.

Control input generator C_i. The control input generator C_i determines the local control input $\boldsymbol{u}_i(t)$ using the model

$$C_i : \begin{cases} \Sigma_\text{s} \text{ defined in (5.3)} \\ \boldsymbol{u}_i(t) = -\boldsymbol{K}_i \boldsymbol{x}_\text{s}(t) \end{cases} \tag{5.4}$$

where $\boldsymbol{K}_i \in \mathbb{R}^{m_i \times n}$ is obtained from the overall state-feedback gain $\boldsymbol{K}$ according to (4.4). At the event times t_{k_j}, C_i receives the state information $\boldsymbol{x}_j(t_{k_j})$ which is used to reset the model state $\boldsymbol{x}_\text{s}(t)$ as indicated in (5.3) and to determine a new disturbance estimate $\hat{\boldsymbol{d}}_k$. A method for the disturbance estimation is presented in Sec. 5.2.4.

Event generator E_i. The event generator E_i continuously measures the subsystem state $\boldsymbol{x}_i(t)$ and triggers an event, whenever the condition

$$\|\boldsymbol{x}_i(t) - \boldsymbol{x}_{\text{s}i}(t)\|_\infty < \bar{e}_i \tag{5.5}$$

is violated, where $\boldsymbol{x}_{\text{s}i} \in \mathbb{R}^{n_i}$ denotes the prediction of the corresponding subsystem state $\boldsymbol{x}_i(t)$. Hence, the event times t_{k_i} are determined by E_i described by the model

$$E_i : \begin{cases} \Sigma_\text{s} \text{ defined in (5.3)} \\ t_0 = 0, \\ t_{k_i} := \inf \left\{ t > t_{k-1} \mid \|\boldsymbol{x}_i(t) - \boldsymbol{x}_{\text{s}i}(t)\|_\infty = \bar{e}_i \right\}. \end{cases} \tag{5.6}$$

Whenever E_i or some other event generator E_j $(j \in \mathcal{N}, j \neq i)$ triggers an event, E_i resets the model state $\boldsymbol{x}_\text{s}(t)$ using the current local state $\boldsymbol{x}_i(t_{k_i})$ or the received information $\boldsymbol{x}_j(t_{k_j})$, respectively. In either case, E_i also updates the disturbance estimate according to the method presented in Sec. 5.2.4.

5.2.3 Behavior of the event-based state-feedback loop

The following analysis investigates the behavior of the event-based state-feedback loop for the time $t \in [t_k, t_{k+1})$ in between two consecutive events. It is assumed that at the beginning of this interval $t_k = t_{k_i}$ holds, i.e., the event generator E_i has triggered an event at time t_k, whereas the next event at time t_{k+1} is triggered by another E_j $(j \in \mathcal{N})$.

The overall plant (5.1a), (5.1b) together with the control input generators C_i defined in (5.4) of the local event-based controllers F_i is described for the time $t \in [t_k, t_{k+1})$ by the state-space model

$$\begin{pmatrix} \dot{\boldsymbol{x}}(t) \\ \dot{\boldsymbol{x}}_{\mathrm{s}}(t) \end{pmatrix} = \begin{pmatrix} \boldsymbol{A} & -\boldsymbol{BK} \\ & \bar{\boldsymbol{A}} \end{pmatrix} \begin{pmatrix} \boldsymbol{x}(t) \\ \boldsymbol{x}_{\mathrm{s}}(t) \end{pmatrix} + \begin{pmatrix} \boldsymbol{E} & \\ & \boldsymbol{E} \end{pmatrix} \begin{pmatrix} \boldsymbol{d}(t) \\ \hat{\boldsymbol{d}}_k \end{pmatrix}$$
$$\begin{pmatrix} \boldsymbol{x}(t_k^+) \\ \boldsymbol{x}_{\mathrm{s}}(t_k^+) \end{pmatrix} = \begin{pmatrix} \boldsymbol{I}_n & \\ \boldsymbol{\Gamma}_i & \boldsymbol{I}_n - \boldsymbol{\Gamma}_i \end{pmatrix} \begin{pmatrix} \boldsymbol{x}(t_k) \\ \boldsymbol{x}_{\mathrm{s}}(t_k) \end{pmatrix}$$

with

$$\boldsymbol{\Gamma}_i = \mathrm{diag}\left(\boldsymbol{O}_{n_1}, \ldots, \boldsymbol{O}_{n_{i-1}}, \boldsymbol{I}_{n_i}, \boldsymbol{O}_{n_{i+1}}, \ldots, \boldsymbol{O}_{n_N}\right). \tag{5.7}$$

To analyze the behavior of the system consider the state transformation

$$\begin{pmatrix} \boldsymbol{x}_\Delta(t) \\ \boldsymbol{x}_{\mathrm{s}}(t) \end{pmatrix} = \begin{pmatrix} \boldsymbol{I}_n & -\boldsymbol{I}_n \\ & \boldsymbol{I}_n \end{pmatrix} \begin{pmatrix} \boldsymbol{x}(t) \\ \boldsymbol{x}_{\mathrm{s}}(t) \end{pmatrix}$$

which yields the transformed state-space representation

$$\begin{pmatrix} \dot{\boldsymbol{x}}_\Delta(t) \\ \dot{\boldsymbol{x}}_{\mathrm{s}}(t) \end{pmatrix} = \begin{pmatrix} \boldsymbol{A} & \\ & \bar{\boldsymbol{A}} \end{pmatrix} \begin{pmatrix} \boldsymbol{x}_\Delta(t) \\ \boldsymbol{x}_{\mathrm{s}}(t) \end{pmatrix} + \begin{pmatrix} \boldsymbol{E} & \\ & \boldsymbol{E} \end{pmatrix} \begin{pmatrix} \boldsymbol{d}_\Delta(t) \\ \hat{\boldsymbol{d}}_k \end{pmatrix} \tag{5.8a}$$
$$\begin{pmatrix} \boldsymbol{x}_\Delta(t_k^+) \\ \boldsymbol{x}_{\mathrm{s}}(t_k^+) \end{pmatrix} = \begin{pmatrix} \boldsymbol{I}_n - \boldsymbol{\Gamma}_i & \\ \boldsymbol{\Gamma}_i & \boldsymbol{I}_n \end{pmatrix} \begin{pmatrix} \boldsymbol{x}_\Delta(t_k) \\ \boldsymbol{x}_{\mathrm{s}}(t_k) \end{pmatrix} \tag{5.8b}$$

with the disturbance estimation error $\boldsymbol{d}_\Delta(t) = \boldsymbol{d}(t) - \hat{\boldsymbol{d}}_k$. From the model (5.8) the equations

$$\boldsymbol{x}_\Delta(t) = \mathrm{e}^{\boldsymbol{A}(t-t_k)}(\boldsymbol{I}_n - \boldsymbol{\Gamma}_i)\boldsymbol{x}_\Delta(t_k) + \int_{t_k}^{t} \mathrm{e}^{\boldsymbol{A}(t-\tau)} \boldsymbol{E}\boldsymbol{d}_\Delta(\tau)\mathrm{d}\tau \tag{5.9}$$
$$\boldsymbol{x}_{\mathrm{s}}(t) = \mathrm{e}^{\bar{\boldsymbol{A}}(t-t_k)}(\boldsymbol{x}_{\mathrm{s}}(t_k) + \boldsymbol{\Gamma}_i\boldsymbol{x}_\Delta(t_k)) + \int_{t_k}^{t} \mathrm{e}^{\bar{\boldsymbol{A}}(t-\tau)} \boldsymbol{E}\hat{\boldsymbol{d}}_k\mathrm{d}\tau \tag{5.10}$$

follow. Equations (5.9) and (5.10) emphasize the main difference between the event-based state-feedback approach [115] and its distributed realization by the proposed method. In [115] the model state $\boldsymbol{x}_{\mathrm{s}}(t)$ is completely reset to the actual plant state $\boldsymbol{x}(t_k)$ at the event time t_k, whereas in the distributed realization only a part of the model state is updated with the current subsystem state, depending on which event generator has triggered the event. This difference

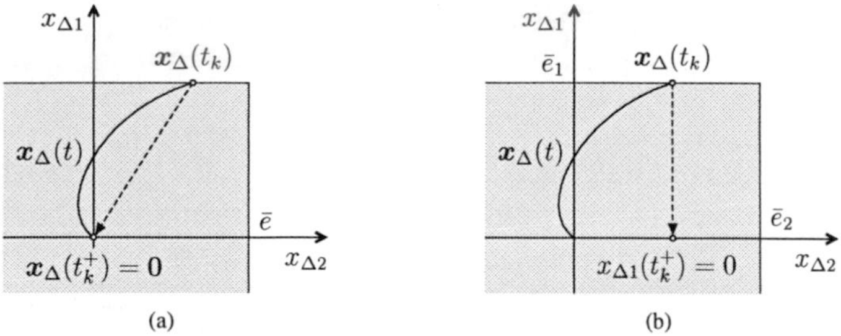

Figure 5.3: Reset of the model state $\boldsymbol{x}_s$ (a) in the event-based state-feedback approach [115] and (b) in the distributed realization

is illustrated in Fig. 5.3 for the example of a systems that is composed of two interconnected scalar subsystems, where the trajectories of the difference state $\boldsymbol{x}_\Delta(t)$ for $t \geq t_{k-1}$ is shown for both methods. In this figure the situation is considered where $\boldsymbol{x}_\Delta(t_{k-1}) = 0$ holds. The identical initial condition implies that both systems behave the same until an event occurs at time t_k when $\|\boldsymbol{x}_\Delta\|_\infty = \bar{e}$ holds in the event-based state-feedback approach (Fig. 5.3(a)) or when $\|x_{\Delta 1}\|_\infty = \|\boldsymbol{\Gamma}_i \boldsymbol{x}_\Delta(t)\|_\infty = \bar{e}_1$ holds in the distributed realization (Fig. 5.3(b)). In the first case the complete difference state $\boldsymbol{x}_\Delta(t)$ is set to zero whereas in the latter case only the component $x_{\Delta 1}(t)$ of $\boldsymbol{x}_\Delta(t)$ is reset. Consequently the behavior of both systems is different for $t \geq t_k$. This consideration leads to the following analysis problem:

> How can the distributed realization of the event-based state feedback be made to approximate the behavior of the reference system (5.2) with the same precision that would be obtained by the centralized event-based state-feedback approach [115]?

An answer to this question is given in Sec. 5.2.5.

5.2.4 Disturbance estimation

The disturbance estimation method presented in Sec. 3.2.3 requires the complete plant state $\boldsymbol{x}(t_k)$ to be given at the time t_k when the new estimate $\hat{\boldsymbol{d}}_k$ is calculated. This assumption is not satisfied for the distributed realization of the event-based state feedback, because at the event time t_{k_i} the state $\boldsymbol{x}_i(t_{k_i})$ is the only current information that is known to all components of the event-based controller.

In the following, the disturbance estimation method presented in 3.2.3 is extended in order to make it applicable to a networked controller F with the structure introduced in Sec. 4.1. This estimation method seizes the idea that the disturbance $\boldsymbol{d}(t)$ is constant

$$\boldsymbol{d}(t) = \boldsymbol{d}_{\mathrm{c}}, \quad \text{for } t \in [t_{k-1}, t_k)$$

where $\boldsymbol{d}_{\mathrm{c}} \in \mathcal{D}$ is the disturbance magnitude and t_{k-1} and t_k denote two consecutive events, possibly triggered by two different event generators. Consider the difference

$$\boldsymbol{x}_{\Delta}(t) := \boldsymbol{x}(t) - \boldsymbol{x}_{\mathrm{s}}(t).$$

According to (5.8), $\boldsymbol{x}_{\Delta}(t)$ is described for $t \in [t_{k-1}, t_k)$ by the state-space model

$$\dot{\boldsymbol{x}}_{\Delta}(t) = \boldsymbol{A}\boldsymbol{x}_{\Delta}(t) + \boldsymbol{E}\big(\boldsymbol{d}_{\mathrm{c}} - \hat{\boldsymbol{d}}_{k-1}\big), \qquad (5.11a)$$

$$\boldsymbol{x}_{\Delta}(t_{k-1}^{+}) = \big(\boldsymbol{I}_n - \boldsymbol{\Gamma}_i\big)\boldsymbol{x}_{\Delta}(t_{k-1}) \qquad (5.11b)$$

with $\boldsymbol{\Gamma}_i$ defined in (5.7). Equation (5.11) yields

$$\begin{aligned} \boldsymbol{x}_{\Delta}(t) &= \breve{\boldsymbol{x}}_{\Delta}(t) + \int_{t_{k-1}}^{t} \mathrm{e}^{\boldsymbol{A}(t-\tau)}\boldsymbol{E}\left(\boldsymbol{d}_{\mathrm{c}} - \hat{\boldsymbol{d}}_{k-1}\right)\mathrm{d}\tau \\ &= \breve{\boldsymbol{x}}_{\Delta}(t) + \boldsymbol{A}^{-1}\left(\mathrm{e}^{\boldsymbol{A}(t-t_{k-1})} - \boldsymbol{I}_n\right)\boldsymbol{E}\left(\boldsymbol{d}_{\mathrm{c}} - \hat{\boldsymbol{d}}_{k-1}\right) \end{aligned} \qquad (5.12)$$

where

$$\breve{\boldsymbol{x}}_{\Delta}(t) = \mathrm{e}^{\boldsymbol{A}(t-t_{k-1})}\big(\boldsymbol{I}_n - \boldsymbol{\Gamma}_i\big)\boldsymbol{x}_{\Delta}(t_{k-1}) \qquad (5.13)$$

denotes the free motion of the system (5.11) which occurs due to the non-vanishing initial condition $\boldsymbol{x}_{\Delta}(t_{k-1})$. Now consider that at the next event time $t = t_k$, Eq. (5.12) results in

$$\boldsymbol{x}_{\Delta}(t_k) = \breve{\boldsymbol{x}}_{\Delta}(t_k) + \boldsymbol{A}^{-1}\left(\mathrm{e}^{\boldsymbol{A}(t_k-t_{k-1})} - \boldsymbol{I}_n\right)\boldsymbol{E}\left(\boldsymbol{d}_{\mathrm{c}} - \hat{\boldsymbol{d}}_{k-1}\right)$$

which is used to determine the new disturbance estimate. That constant disturbance $\boldsymbol{d}_{\mathrm{c}}$ which solves the last equation is taken as the new disturbance estimate $\hat{\boldsymbol{d}}_k$ for $t \geq t_k$. Under the condition that the dimension p of the disturbance $\boldsymbol{d}(t)$ is not larger than the dimension n of the state $\boldsymbol{x}(t)$ and that the matrix

$$\left(\boldsymbol{A}^{-1}\left(\mathrm{e}^{\boldsymbol{A}(t_k-t_{k-1})} - \boldsymbol{I}_n\right)\boldsymbol{E}\right)^{\top}\left(\boldsymbol{A}^{-1}\left(\mathrm{e}^{\boldsymbol{A}(t_k-t_{k-1})} - \boldsymbol{I}_n\right)\boldsymbol{E}\right)$$

has full rank, this equation has the explicit solution

$$\hat{\boldsymbol{d}}_k := \boldsymbol{d}_{\mathrm{c}} = \hat{\boldsymbol{d}}_{k-1} + \left(\boldsymbol{A}^{-1}\left(\mathrm{e}^{\boldsymbol{A}(t_k-t_{k-1})} - \boldsymbol{I}_n\right)\boldsymbol{E}\right)^{+} \cdot \left(\boldsymbol{x}_{\Delta}(t_k) - \breve{\boldsymbol{x}}_{\Delta}(t_k)\right). \qquad (5.14)$$

Note that the states $\boldsymbol{x}_\Delta(t_{k-1})$ and $\boldsymbol{x}_\Delta(t_k)$ need to be known to evaluate the previous equation. However, in the control input generators and the event generators the overall difference state $\boldsymbol{x}_\Delta(t)$ is not accessible. The only information about the difference state $\boldsymbol{x}_\Delta$ that is known to all components at some event time $t_k = t_{k_i}$ is

$$\boldsymbol{x}_{\Delta i}(t_{k_i}) = \boldsymbol{x}_i(t_{k_i}) - \boldsymbol{x}_{\mathrm{s}i}(t_{k_i}),$$

because $\boldsymbol{x}_i(t_{k_i})$ is broadcasted by E_i through the network. In order to determine the estimate $\hat{\boldsymbol{d}}_k$ in all components the assumption

$$\boldsymbol{x}_j(t_{k_i}) = \boldsymbol{x}_{\mathrm{s}j}(t_{k_i}), \quad \forall\, j \in \mathcal{N} \setminus \{i\}$$

is made. This implies that at the event time $t_k = t_{k_i}$ the approximation

$$\boldsymbol{\Gamma}_i \boldsymbol{x}_\Delta(t_{k_i}) = \begin{pmatrix} \boldsymbol{0}_{n_1}^\top & \dots & \boldsymbol{0}_{n_{i-1}}^\top & (\boldsymbol{x}_i(t_{k_i}) - \boldsymbol{x}_{\mathrm{s}i}(t_{k_i}))^\top & \boldsymbol{0}_{n_{i+1}}^\top & \dots & \boldsymbol{0}_N^\top \end{pmatrix}^\top$$

is applied to the estimation (5.14) instead of the actual difference state $\boldsymbol{x}_\Delta(t_k)$. By this means the free motion (5.13) vanishes and the estimation (5.14) reduces to

$$\hat{\boldsymbol{d}}_0 = \boldsymbol{0}, \tag{5.15a}$$

$$\hat{\boldsymbol{d}}_k = \hat{\boldsymbol{d}}_{k-1} + \left(\boldsymbol{A}^{-1} \left(\mathrm{e}^{\boldsymbol{A}(t_{k_i} - t_{k-1})} - \boldsymbol{I}_n \right) \boldsymbol{E} \right)^+ \boldsymbol{\Gamma}_i \boldsymbol{x}_\Delta(t_{k_i}) \tag{5.15b}$$

where the initial estimate $\hat{\boldsymbol{d}}_0$ is set to zero if no other information is given about the disturbance $\boldsymbol{d}(t)$. Thus, the estimation method (5.15) differs from the method (3.6) in that the approximation $\boldsymbol{\Gamma}_i \boldsymbol{x}_\Delta(t_k)$ instead of the actual difference state $\boldsymbol{x}_\Delta(t_k)$ is applied, which is known to all event generators and control input generators at the event time t_k. This means that on the one hand the estimation method (5.15) can be implemented in each control input generator and each event generator and on the other hand all generators determine the same estimates $\hat{\boldsymbol{d}}_k$ for all $k \in \mathbb{N}_0$. Particularly the second property is crucial for the proposed distributed realization of the event-based state feedback as this method is based on the requirement that the models Σ_{s} given in (5.3) are synchronized for all $t \geq 0$ in all control input generators and all event generators in order to have the same prediction $\boldsymbol{x}_{\mathrm{s}}(t)$ of the pant state $\boldsymbol{x}(t)$.

5.2.5 Approximation of the reference system behavior

This section investigates the deviation between the behavior of the distributed realization of the event-based state feedback (5.1a), (5.1b), (5.4), (5.6) and the reference system (5.2).

First, note that the event condition (5.5) that is applied by each event generator E_i ($i \in \mathcal{N}$)

implies

$$\|\boldsymbol{x}_\Delta(t)\|_\infty = \|\boldsymbol{x}(t) - \boldsymbol{x}_\mathrm{s}(t)\|_\infty \le \max(\bar{e}_1, \dots, \bar{e}_N), \quad \forall\, t \ge 0. \tag{5.16}$$

Hence, the relation

$$\|\boldsymbol{K}_i\left(\boldsymbol{x}(t) - \boldsymbol{x}_\mathrm{s}(t)\right)\|_\infty \le \|\boldsymbol{K}_i\|_\infty \|\boldsymbol{x}_\Delta(t)\|_\infty \le \|\boldsymbol{K}_i\|_\infty \max(\bar{e}_1, \dots, \bar{e}_N)$$

holds for all $t \ge 0$, which shows that the condition (4.5) is satisfied. Thus, according to Theorem 4.1, the deviation between the behavior of the distributed realization of the event-based state feedback (5.1a), (5.1b), (5.4), (5.6) and the reference system (5.2) is bounded. The following analysis makes this result more precise and determines an upper bound for this deviation.

Given that the models Σ_s are synchronized in all generators, from Eq. (4.7) the model

$$\dot{\boldsymbol{e}}(t) = \bar{\boldsymbol{A}}\boldsymbol{e}(t) + \boldsymbol{B}\boldsymbol{K}\boldsymbol{x}_\Delta(t), \quad \boldsymbol{e}(0) = \boldsymbol{0} \tag{5.17}$$

follows, where $\boldsymbol{e}(t) = \boldsymbol{x}(t) - \boldsymbol{x}_\mathrm{r}(t)$ denotes the deviation between the plant state $\boldsymbol{x}(t)$ in distributed realization of the event-based state feedback and the reference system state $\boldsymbol{x}_\mathrm{r}(t)$. The last equation yields

$$\begin{aligned}\|\boldsymbol{e}(t)\| &= \left\| \int_0^t \mathrm{e}^{\bar{\boldsymbol{A}}(t-\tau)} \boldsymbol{B}\boldsymbol{K}\boldsymbol{x}_\Delta(\tau)\mathrm{d}\tau \right\|, \\ &\le \int_0^t \left\| \mathrm{e}^{\bar{\boldsymbol{A}}(t-\tau)} \boldsymbol{B}\boldsymbol{K} \right\| \|\boldsymbol{x}_\Delta(\tau)\| \,\mathrm{d}\tau.\end{aligned}$$

Since the matrix $\bar{\boldsymbol{A}}$ is Hurwitz by design and the difference state $\boldsymbol{x}_\Delta(t)$ is bounded as stated in Eq. (5.16), the approximation error $\boldsymbol{e}(t)$ is bounded as well which leads to the following theorem.

Theorem 5.1 *The approximation error $\boldsymbol{e}(t) = \boldsymbol{x}(t) - \boldsymbol{x}_\mathrm{r}(t)$ between the state of the distributed realization of the event-based state feedback* (5.1a), (5.1b), (5.4), (5.6) *and the state of the reference system* (5.2) *is bounded from above by*

$$\|\boldsymbol{e}(t)\|_\infty \le \max(\bar{e}_1, \dots, \bar{e}_N) \cdot \int_0^\infty \left\| \mathrm{e}^{\bar{\boldsymbol{A}}\tau} \boldsymbol{B}\boldsymbol{K} \right\|_\infty \mathrm{d}\tau =: e_\mathrm{max} \tag{5.18}$$

for all $t \ge 0$.

The distributed realization of the event-based state feedback can be made to approximate the behavior of the reference system (5.2) with arbitrary accuracy, because the bound on the error depends linearly on the maximum event threshold. A comparison of Eqs. (3.7) and (5.18) shows that the distributed realization of the event-based state feedback with

$$\max(\bar{e}_1, \ldots, \bar{e}_N) = \bar{e}$$

yields the same maximum deviation to the reference system (5.2) which would be obtained by the centralized event-based state-feedback approach where the event threshold $\bar{e}$ is applied.

Note that Theorem 5.1 can be viewed as a proof for the stability of the event-based control system (5.1a), (5.1b), (5.4), (5.6), since the state $\boldsymbol{x}_\mathrm{r}(t)$ of the reference system (5.2) is practically stable with respect to the set $\mathcal{A}_\mathrm{r}$ given in (3.2). Hence, the event-based control system is practically stable with respect to the set

$$\mathcal{A} := \left\{ \boldsymbol{x} \in \mathrm{I\!R}^n \;\middle|\; \|\boldsymbol{x}\| \leq b_\mathrm{r} + e_{\max} \right\}$$

with the ultimate bound b_r given in (3.2).

An important inference of this analysis is that the stability of the event-based control system is independent of the coupling strength for the interconnection between the subsystems.

> Given that the reference system (5.2) is practically stable, the distributed realization of the event-based state-feedback approach with the control input generators C_i and the event generators E_i as described in (5.4) and (5.6), respectively, ensures the practical stability of the overall control system regardless of the strength of the intercnnection (5.1b) between the subsystems (5.1a).

5.2.6 Minimum inter-event time

This section presents a method for the analysis of the minimum time that elapses in between consecutive events triggered by one event generator E_i for some $i \in \mathcal{N}$, referred to as the minimum inter-event time

$$T_{\min i} := \min_{k_i} \left(t_{k_i+1} - t_{k_i} \right), \quad \forall\, k_i \in \mathrm{I\!N}_0.$$

The following Theorem shows that the minimum inter-event time $T_{\min i}$ is bounded from below by some time $\bar{T}_i$.

Theorem 5.2 *Consider the interconnected subsystems (5.1a), (5.1b) together with the event-based controllers (5.4), (5.6). The minimum inter-event time $T_{\min i}$ for two consecutive events triggered by the event generator E_i is bounded from below by*

$$\bar{T}_i = \arg\min_t \left\{ \int_0^t \left\| \mathrm{e}^{\boldsymbol{A}_i \tau} \right\|_\infty \mathrm{d}\tau = \frac{\bar{e}_i}{\|\boldsymbol{E}_i\|_\infty \bar{d}_{\Delta i} + \sum_{j=1}^N \|\boldsymbol{E}_{si} \boldsymbol{L}_{ij} \boldsymbol{C}_{zj}\|_\infty \bar{e}_j} \right\} \tag{5.19}$$

where $\bar{d}_{\Delta i} \geq \left\| \boldsymbol{d}_i(t) - \hat{\boldsymbol{d}}_{ik} \right\|_\infty$ denotes the maximum deviation between the local disturbance $\boldsymbol{d}_i(t)$ and the i-th component $\hat{\boldsymbol{d}}_{ik}$ of the disturbance estimate $\hat{\boldsymbol{d}}_k$ for all $t \geq 0$ and all $k \in \mathbb{N}_0$.

Before the proof of Theorem 5.2 is given, the result (5.19) is interpreted first. An interesting fact of (5.19) is that the bound $\bar{T}_i$ on the minimum inter-event time $T_{\min i}$ in between two events triggered by event generator E_i does not only depend upon the event threshold $\bar{e}_i$ but also upon the event thresholds $\bar{e}_j$ of all other event generators E_j $(j \in \mathcal{N})$. Hence, the choice of the local event threshold $\bar{e}_i$ affects the bound $\bar{T}_i$ as well as the bounds $\bar{T}_j$ on the minimum inter-event times $T_{\min j}$. More precisely, the increase of the event threshold $\bar{e}_i$ increases the inter-event time $T_{\min i}$ but at the same time decreases the inter-event times $T_{\min j}$ for events triggered by E_j for all $j \in \mathcal{N}$, $j \neq i$. This means that a more frequent event triggering by the event generators E_j due to smaller event thresholds $\bar{e}_j$ postpones the triggering of events generated by E_i. Besides the event thresholds the minimum inter-event time $T_{\min i}$ also depends upon the local disturbance $\boldsymbol{d}_i(t)$. Similar to the bound (3.10) for the inter-event time in the centralized event-based state-feedback approach, the event generator E_i triggers events more frequently if the actual disturbance $\boldsymbol{d}_i(t)$ deviates more from the estimate $\hat{\boldsymbol{d}}_{ik}$ and vice versa.

Proof. Consider the subsystem Σ_i together with the control input generator C_i, represented by Eqs. (5.1a) and (5.4), respectively:

$$\dot{\boldsymbol{x}}_i(t) = \boldsymbol{A}_i \boldsymbol{x}_i(t) - \boldsymbol{B}_i \boldsymbol{K}_i \boldsymbol{x}_s(t) + \boldsymbol{E}_i \boldsymbol{d}_i(t) + \boldsymbol{E}_{si} \boldsymbol{s}_i(t), \quad \boldsymbol{x}_i(0) = \boldsymbol{x}_{0i}. \tag{5.20}$$

Now consider the i-th component $\boldsymbol{x}_{si}(t)$ of the overall model state $\boldsymbol{x}_s(t)$ which is described in the interval $[t_{k_i}, t_{k+1})$ by the state-space model

$$\dot{\boldsymbol{x}}_{si}(t) = \boldsymbol{A}_i \boldsymbol{x}_{si}(t) - \boldsymbol{B}_i \boldsymbol{K}_i \boldsymbol{x}_s(t) + \boldsymbol{E}_i \hat{\boldsymbol{d}}_{ik} + \boldsymbol{E}_{si} \boldsymbol{s}_{si}(t), \quad \boldsymbol{x}_{si}(t_{k_i}^+) = \boldsymbol{x}_i(t_{k_i}) \tag{5.21a}$$

$$\boldsymbol{z}_{si}(t) = \boldsymbol{C}_{zi} \boldsymbol{x}_{si}(t), \tag{5.21b}$$

where

$$\boldsymbol{s}_{si}(t) = \sum_{j=1}^{N} \boldsymbol{L}_{ij} \boldsymbol{z}_{sj}(t) = \sum_{j=1}^{N} \boldsymbol{L}_{ij} \boldsymbol{C}_{zj} \boldsymbol{x}_{sj}(t). \tag{5.22}$$

With $\boldsymbol{x}_{\Delta i}(t) = \boldsymbol{x}_i(t) - \boldsymbol{x}_{si}(t)$, Eqs. (5.1b), (5.20)–(5.22) yield

$$\begin{aligned}
\dot{\boldsymbol{x}}_{\Delta i}(t) &= \boldsymbol{A}_i \boldsymbol{x}_{\Delta i}(t) + \boldsymbol{E}_i\big(\boldsymbol{d}_i(t) - \hat{\boldsymbol{d}}_{ik}\big) + \boldsymbol{E}_{si} \boldsymbol{s}_{\Delta i}(t), \quad \boldsymbol{x}_{\Delta i}(t_{k_i}^+) = \boldsymbol{0} \\
\boldsymbol{s}_{\Delta i}(t) &= \boldsymbol{s}_i(t) - \boldsymbol{s}_{si}(t) = \sum_{j=1}^{N} \boldsymbol{L}_{ij} \boldsymbol{C}_{zj} \boldsymbol{x}_{\Delta j}(t)
\end{aligned}$$

which describes the behavior of the difference state $\boldsymbol{x}_{\Delta i}(t)$ for the time $t \in [t_{k_i}, t_{k+1})$ and results in

$$\boldsymbol{x}_{\Delta i}(t) = \int_{t_{k_i}}^{t} \mathrm{e}^{\boldsymbol{A}_i(t-\tau)} \left(\boldsymbol{E}_i\big(\boldsymbol{d}_i(\tau) - \hat{\boldsymbol{d}}_{ik}\big) + \sum_{j=1}^{N} \boldsymbol{E}_{si} \boldsymbol{L}_{ij} \boldsymbol{C}_{zj} \boldsymbol{x}_{\Delta j}(\tau) \right) \mathrm{d}\tau.$$

Assume that the next event is triggered at time $t_{k+1} = t_{k_j}$ by the event generator E_j. At the event time t_{k_j} only the difference state $\boldsymbol{x}_{\Delta j}$ is reset to zero, whereas for the difference state $\boldsymbol{x}_{\Delta i}$ the relation

$$\boldsymbol{x}_{\Delta i}(t_{k_j}^+) = \boldsymbol{x}_{\Delta i}(t_{k_j}) \tag{5.23}$$

holds. Hence, for the time $t \in [t_{k_j}, t_{k+2})$, the difference state $\boldsymbol{x}_{\Delta i}(t)$ is given by

$$\begin{aligned}
\boldsymbol{x}_{\Delta i}(t) = \mathrm{e}^{\boldsymbol{A}_i(t-t_{k_j})} \boldsymbol{x}_{\Delta i}(t_{k_j}) \\
+ \int_{t_{k_j}}^{t} \mathrm{e}^{\boldsymbol{A}_i(t-\tau)} \left(\boldsymbol{E}_i\big(\boldsymbol{d}_i(\tau) - \hat{\boldsymbol{d}}_{i(k+1)}\big) + \sum_{j=1}^{N} \boldsymbol{E}_{si} \boldsymbol{L}_{ij} \boldsymbol{C}_{zj} \boldsymbol{x}_{\Delta j}(\tau) \right) \mathrm{d}\tau
\end{aligned}$$

By virtue of (5.23), for the time $t \in [t_{k_i}, t_{k+2})$ the difference state $\boldsymbol{x}_{\Delta i}(t)$ is described by the equation

$$\boldsymbol{x}_{\Delta i}(t) = \int_{t_{k_i}}^{t} \mathrm{e}^{\boldsymbol{A}_i(t-\tau)} \left(\boldsymbol{E}_i \boldsymbol{d}_{\Delta i}(\tau) + \sum_{j=1}^{N} \boldsymbol{E}_{si} \boldsymbol{L}_{ij} \boldsymbol{C}_{zj} \boldsymbol{x}_{\Delta j}(\tau) \right) \mathrm{d}\tau, \tag{5.24}$$

where $\boldsymbol{d}_{\Delta i}(t) = \boldsymbol{d}_i(t) - \hat{\boldsymbol{d}}_{ik}$ denotes the deviation between the actual disturbance $\boldsymbol{d}_i(t)$ and its estimate $\hat{\boldsymbol{d}}_{ik}$ for all $t \geq 0$ and all $k \in \mathbb{N}_0$. Now consider the sequence of event times $\{t_{k_i}, t_{k+1}, \ldots, t_{k+\kappa}\}$ for some $\kappa \in \mathbb{N}$, $\kappa > 1$ and assume that t_{k_i} denotes the last time at which E_i has triggered an event. Following the same argumentation as before it can be concluded that

Eq. (5.24) also holds for the time $t \in [t_{k_i}, t_{k+\kappa})$.

Equation (5.24) is applied in the following analysis to determine a bound $\bar{T}_i$ on the minimum inter-event time $T_{\min i}$. Consider that the next event is triggered by E_i at time t_{k_i+1} at which

$$\|\boldsymbol{x}_{\Delta i}(t)\|_\infty = \left\| \int_{t_{k_i}}^{t} \mathrm{e}^{\boldsymbol{A}_i(t-\tau)} \left(\boldsymbol{E}_i \boldsymbol{d}_{\Delta i}(\tau) + \sum_{j=1}^{N} \boldsymbol{E}_{si} \boldsymbol{L}_{ij} \boldsymbol{C}_{zj} \boldsymbol{x}_{\Delta j}(\tau) \right) \mathrm{d}\tau \right\|_\infty = \bar{e}_i$$

holds. In order to find a lower bound $\bar{T}_i$ on $T_{\min i}$ the estimate

$$\|\boldsymbol{x}_{\Delta i}(t)\|_\infty \leq \int_{t_{k_i}}^{t} \left\| \mathrm{e}^{\boldsymbol{A}_i \tau} \right\|_\infty \left(\|\boldsymbol{E}_i\|_\infty \bar{d}_{\Delta i} + \sum_{j=1}^{N} \|\boldsymbol{E}_{si} \boldsymbol{L}_{ij} \boldsymbol{C}_{zj}\|_\infty \bar{e}_j \right)$$

is used, where $\bar{d}_{\Delta i} \geq \|\boldsymbol{d}_{\Delta i}(t)\|_\infty$ holds for all $t \geq 0$. Finally,

$$\bar{T}_i = \arg\min_t \left\{ \int_{t_{k_i}}^{t} \left\| \mathrm{e}^{\boldsymbol{A}_i \tau} \right\|_\infty \left(\|\boldsymbol{E}_i\|_\infty \bar{d}_{\Delta i} + \sum_{j=1}^{N} \|\boldsymbol{E}_{si} \boldsymbol{L}_{ij} \boldsymbol{C}_{zj}\|_\infty \bar{e}_j \right) = \bar{e}_i \right\}$$

represents a lower bound on $T_{\min i}$, which completes the proof. □

Note that the minimum inter-event time for two consecutive events that are triggered by two different event generators E_i and E_j $(i, j \in \mathcal{N}, i \neq j)$ is zero, because the triggering condition (5.5) can be violated in these generators at the same time. In view of the fact that the inter-event time for each event generator E_i is bounded from below by (5.19), at the most N events are triggered at a time. This implies that in finite time only a finite number of events are triggered and, hence, Zeno behavior does not occur in the distributed realization of the event-based state feedback.

5.2.7 Discussion of the control approach

The analysis in Sec. 5.2.5 has shown that the approach to the distributed realization of the event-based state feedback can be made to yield the same performance with respect to the approximation of the reference system behavior as the centralized event-based state feedback would yield for the system with centralized sensor and actuator units. In this respect the proposed control approach is a further development of the basic concept for controlling interconnected systems (5.1a), (5.1b). Besides the property that both approaches approximate a desired behavior with adjustable accuracy, there are some differences between the event-based state feedback and the distributed realization which are discussed next.

The first and most obvious difference between both approaches results from the different structures of the plants to be controlled. In the event-based state-feedback approach the plant is assumed to have a central sensor unit and a central actuator unit and, thus, the controller consists of two components (control input generator C and event generator E). As opposed to this, the distributed realization introduces $2N$ components, namely a control input generator and an event generator for each subsystem, where each generator evaluates a model of the overall control system. Hence, in the latter control approach the computational effort is increased considerably compared to the centralized event-based state feedback. On the other hand, the distributed realization of the event triggering brings along a higher flexibility for the weighting of the states in the triggering condition. In the centralized approach the event times are determined based on one event condition only, whereas in the distributed realization the triggering condition is checked locally and for each event generator a separate event threshold can be chosen.

An important property of the distributed realization of the event-based state-feedback approach is that this method is applicable to interconnected systems (5.1a), (5.1b) with arbitrary coupling strength. Theorem 5.1 has shown that the practical stability of the reference system (5.2) with a centralized state feedback implies the practical stability of the event-based control system (5.1a), (5.1b), (5.4), (5.6) independently of the interconnection between the subsystems.

5.3 Example: Event-based state-feedback control of the thermofluid process

This section investigates the behavior of the distributed realization of the event-based state feedback for the example of the thermofluid process introduced in Sec. 2.4 by means of a simulation and an experiment. The simulation results are compared with the results obtained for the centralized event-based state feedback which are presented in Sec. 3.4. The experimental results show that the distributed realization of the event-based state feedback is robust with respect to model uncertainties.

The scenario which is investigated in this section is identical with the one described in Sec. 3.4 and is specified by the following disturbance characteristics:

$$d_1(t) = \begin{cases} 0.1, & \text{if } t \in [200, 600]\,\mathrm{s} \\ 0, & \text{else,} \end{cases}$$

$$d_2(t) = \begin{cases} 0.25, & \text{if } t \in [1000, 1400]\,\mathrm{s} \\ 0, & \text{else.} \end{cases}$$

The focus of this investigation is on the approximation of the behavior of the continuous state-feedback loop with the state-feedback gain given in (3.11) by the distributed realization of the event-based state feedback. Hereafter, the thermofluid process is considered to be composed of two interconnected subsystems, described in Appendix A.2. Hence, the overall event-based controller consists of a control input generator C_i and an event generator E_i for each subsystem Σ_i $(i = 1, 2)$. The event thresholds are set to

$$\bar{e}_1 = 1, \quad \bar{e}_2 = 2,$$

such that $\max(\bar{e}_1, \bar{e}_2) = 2$ equals the event threshold $\bar{e} = 2$ that is chosen for the investigation of the event-based state-feedback approach presented in Sec. 3.4. For the evaluation of the triggering conditions the levels $l_{\mathrm{B}}(t)$ and $l_{\mathrm{S}}(t)$ are again converted into cm. An event is triggered by the event generator E_1 whenever the equality

$$\left\| \begin{pmatrix} 100 \cdot (l_{\mathrm{B}}(t) - l_{\mathrm{sB}}(t)) \\ \vartheta_{\mathrm{B}}(t) - \vartheta_{\mathrm{sB}}(t) \end{pmatrix} \right\|_\infty = \bar{e}_1 = 1$$

holds and, accordingly, E_2 generates an event whenever

$$\left\| \begin{pmatrix} 100 \cdot (l_{\mathrm{S}}(t) - l_{\mathrm{sS}}(t) \\ \vartheta_{\mathrm{S}}(t) - \vartheta_{\mathrm{sS}}(t) \end{pmatrix} \right\|_{\infty} = \bar{e}_2 = 2$$

is satisfied. Here,

$$\boldsymbol{x}_{\mathrm{s}}(t) = \begin{pmatrix} \boldsymbol{x}_{\mathrm{s1}}(t) \\ \boldsymbol{x}_{\mathrm{s2}}(t) \end{pmatrix} = \begin{pmatrix} \left(l_{\mathrm{sB}}(t) \quad \vartheta_{\mathrm{sB}}(t)\right)^{\top} \\ \left(l_{\mathrm{sS}}(t) \quad \vartheta_{\mathrm{sS}}(t)\right)^{\top} \end{pmatrix}$$

denotes the state of the model Σ_{s} given in (5.3).

The analysis of the maximum approximation error $\boldsymbol{e}(t)$ according to Theorem 5.1 yields the bound

$$e_{\max} = 12.01.$$

This bound is identical with the bound that is obtained for the event-based state-feedback approach in the investigation presented in Sec. 3.4, because the event thresholds are chosen such that $\max(\bar{e}_1, \bar{e}_2) = \bar{e}$ holds. With $\bar{d}_{\Delta 1} = 0.2$ and $\bar{d}_{\Delta 2} = 0.5$, Eq. (5.19) yields the bounds

$$\bar{T}_1 = 7\,\mathrm{s}, \qquad \bar{T}_2 = 23\,\mathrm{s}$$

on the minim inter-event times $T_{\min 1}$ and $T_{\min 2}$, respectively.

Simulation results. The disturbance behavior of the distributed realization of the event-based state feedback is illustrated in Fig. 5.4. First, consider the left-hand side of Fig. 5.4 which shows the behavior of reactor TB. In total, 5 events are triggered as reaction to the disturbance $d_1(t)$. After the first four events the disturbance estimate $\hat{d}_{1k}$ deviates from the actual disturbance $d_1(t)$ which is due to the fact that the estimation is based on the assumption $\boldsymbol{x}_{\Delta 2}(t_{k_1}) = \boldsymbol{0}$ (for $k_1 = 1, \ldots, 4$) which, however, does not hold. After the fifth event at time $t_{15} = 781\,\mathrm{s}$ both the level $l_{\mathrm{B}}(t)$ and the temperature $\vartheta_{\mathrm{B}}(t)$ remain close to the setpoint and no more event is triggered. The absence of further events has two reasons: First, reactor TB is no more disturbed and the disturbance estimate only marginally deviates from the actual disturbance. Second, the perturbation due to the couplings to reactor TS is well attenuated by the controller F_1. Note that the attenuation of the coupling input is promoted by the triggering of events of the event generator E_2 which causes the update of the state predictions $\boldsymbol{x}_{\mathrm{s2}}(t)$ also in the models used in the controller F_1.

Now consider the behavior of the reactor TS shown on the right-hand side of Fig. 5.4. In the time interval $t \in [0, 1000]$ s five events are triggered, although reactor TS is not disturbed. The triggering of these events is mainly caused by the disturbance estimates which significantly deviate from the disturbance $d_2(t) = 0$ in this interval. In relation to this, the interconnection to reactor TB has a minor effect on the event triggering. In the time interval $t \in [1000, 1400]$ s the disturbance $d_2(t) = 0.25$ affects reactor TS. In this interval and afterwards the reactor TS is barely affected by the coupling to reactor TB and, thus, the disturbance is accurately estimated. In total, only 9 events are triggered within 2000 s.

In this investigation the minimum inter-event times for both subsystems are

$$T_{\min 1} = 83\,\mathrm{s}, \qquad T_{\min 2} = 34\,\mathrm{s}.$$

The time span $T_{\min 1}$ is by a factor of more than 11 larger than the bound $\bar{T}_1 = 7$ s, whereas the bound $\bar{T}_2 = 23$ s is a much less conservative estimate on the actual minimum inter-event time $T_{\min 2}$.

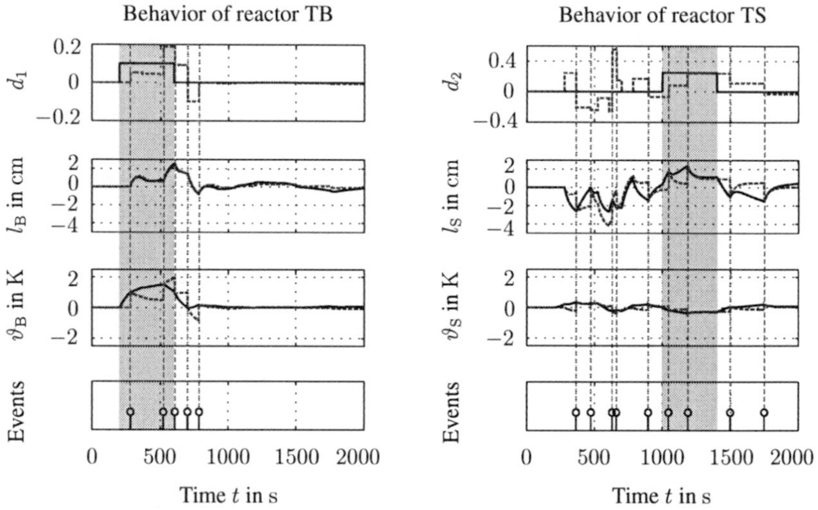

Figure 5.4: Disturbance behavior of the distributed realization of the event-based state-feedback approach. The behavior of reactor TB and reactor TS is plotted on the left-hand side or right-hand side, respectively. The first row shows the disturbances (solid lines) and its estimation (dashed lines). The trajectories of the level and temperature are given in the second and third row (solid line: plant state $\boldsymbol{x}(t)$, dashed line: model state $\boldsymbol{x}_s(t)$). The event time instants are represented by stems in the bottom figure.

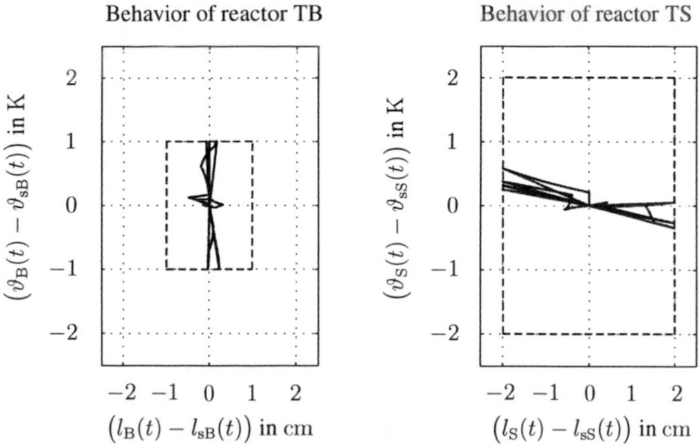

Figure 5.5: Trajectories of the difference states $\boldsymbol{x}_{\Delta 1}(t)$ and $\boldsymbol{x}_{\Delta 2}(t)$.

Figure 5.5 shows the trajectories of the difference states $\boldsymbol{x}_{\Delta 1}(t)$ and $\boldsymbol{x}_{\Delta 2}(t)$. The dashed lines represent the event thresholds $\bar{e}_1 = 1$ and $\bar{e}_2 = 2$. The difference state $\boldsymbol{x}_{\Delta 1}(t)$ or $\boldsymbol{x}_{\Delta 2}(t)$ is reset to zero whenever the corresponding trajectory attains the event threshold $\bar{e}_1$ or $\bar{e}_2$, respectively. Although this reset strategy differs from the one of the event-based state-feedback approach [115], the characteristics of the trajectories in Fig. 5.5 are similar to the trajectories shown in Fig. 3.4. Like in the centralized approach, events are only triggered because of the deviation of the temperature $\vartheta_B(t)$ or the level $l_S(t)$ from the corresponding predictions $\vartheta_{sB}(t)$ or $l_{sS}(t)$, respectively.

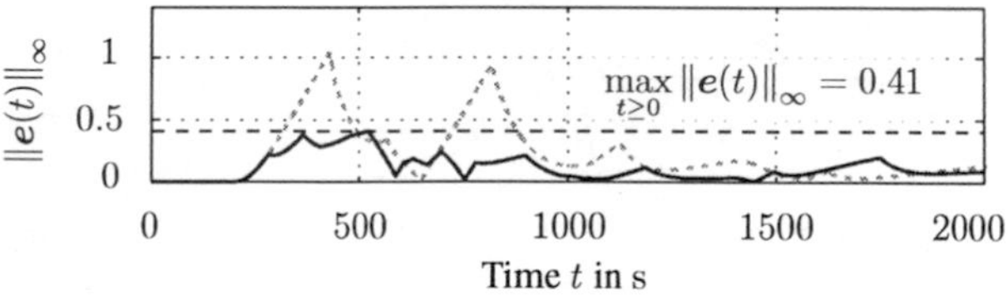

Figure 5.6: Deviation $\boldsymbol{e}(t)$ between the behavior of the event-based state-feedback loop and the reference system.

The deviation between the behavior of the distributed realization of the event-based state feedback and of the continuous state-feedback loop is depicted in Fig. 5.6 by the solid line. The dotted line represents the deviation between the behavior of the centralized event-based state-feedback system and of the same continuous state-feedback loop (cf. Fig. 3.5) which is shown here for a comparison between the centralized approach and its distributed realization.

This figure shows that the maximum approximation error is

$$\max_{t\geq 0} \|e(t)\|_\infty = 0.41$$

and, thus, is less than half the maximum deviation that is observed in the investigation in Sec. 3.4 and, moreover, is considerably smaller than the derived bound $e_{\max} = 12.01$.

Experimental results. Figure 5.7 shows the experimental evaluation of the proposed event-based control method. The obtained results are similar to the simulation results shown in Fig. 5.4, albeit considerably more events are triggered. The fact that more events are triggered compared to the simulation is attributable to model uncertainties. Nevertheless, this investigation shows that the feedback is still adapted to the system behavior. Events are generated more frequently whenever a disturbance is active in the time intervals $t \in [200, 600]$ s and $t \in [1000, 1400]$ s. After the last event triggered by the event generator E_2 at time $t = 1453$ s no event is triggered for more than 500 s. In total 13 events are triggered by the event generator E_1 and 16 by event generator E_2 within the period of 2000 s. Hence, in average every 68 s a

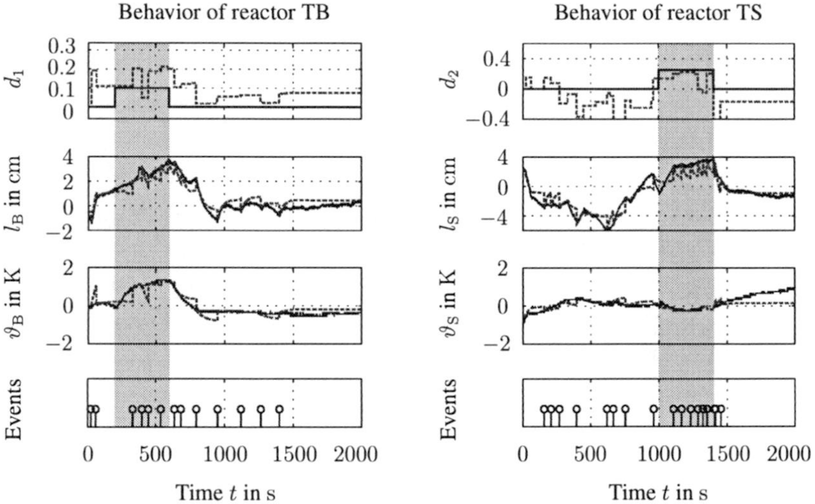

Figure 5.7: Experiemental result of the disturbance behavior of the distributed realization of the event-based state-feedback approach. The behavior of reactor TB and reactor TS is plotted on the left-hand side or right-hand side, respectively. From top to bottom the figures show disturbances (solid lines) and its estimation (dashed lines), the level and temperature (solid line: plant state $x(t)$, dashed line: model state $x_s(t)$) and the event time instants, represented by stems.

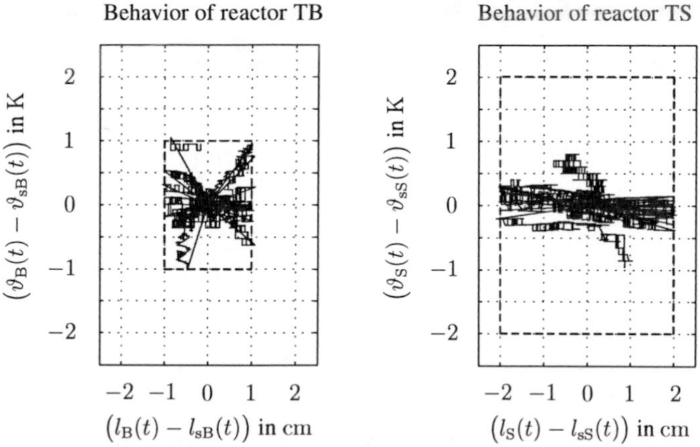

Figure 5.8: Experimental results of the trajectories of the difference states $\boldsymbol{x}_{\Delta 1}(t)$ and $\boldsymbol{x}_{\Delta 2}(t)$.

feedback communication is induced which is a significant reduction compared to a discrete-time control with a sampling period $h = 10\,\mathrm{s}$ (which is an appropriate choice for the considered process).

The trajectories of the difference states $\boldsymbol{x}_{\Delta 1}(t)$ and $\boldsymbol{x}_{\Delta 2}(t)$ are presented in Fig. 5.8. This figure shows that in reactor TB events are caused by a deviation between the temperature $\vartheta_B(t)$ and $\vartheta_{sB}(t)$ as well as by a deviation between the level $l_B(t)$ and $l_{sB}(t)$. This contrasts with the observations that are made in the simulation (cf. Fig. 5.5), where in reactor TB events are only triggered due to a deviation between $\vartheta_B(t)$ and $\vartheta_{sB}(t)$. The reason for the mismatch of the system behavior in the simulation and in the experiment can again be explained by model uncertainties. In reactor TS events are only triggered because of the deviation of the level $l_S(t)$ from the prediction $l_{sS}(t)$ which complies with the simulation (cf. Fig. 5.5).

5.4 Event-based state feedback with incomplete state measurement

The method for the distributed realization of the event-based state feedback proposed in Sec. 5.2 has been shown to be able to approximate the disturbance behavior of the reference system (5.2) with arbitrary accuracy (cf. Theorem 5.1). This property is achieved by appropriately bounding the difference states $\boldsymbol{x}_{\Delta i}(t) = \boldsymbol{x}_i(t) - \boldsymbol{x}_{si}(t)$ for each $i \in \mathcal{N}$ by means of the event triggering which, however, requires all the subsystem states $\boldsymbol{x}_i(t)$ to be measurable. This section presents an extension of the distributed realization of the event-based state feedback where this assumption is removed and plants are considered where the state of only some subsystems is measurable, whereas in the remaining subsystems neither the state nor the output information is available.

The main result of this section is a method for the design of event thresholds $\bar{e}_i$ of those event generators E_i which have access to the measurable states $\boldsymbol{x}_i(t)$. The proposed design method guarantees that a desired disturbance behavior is accomplished by the event-based control system despite the constraints on the state measurement (Theorem 5.3). This event threshold design is formulated as a linear programming problem that is presented in the Algorithm 5.1.

5.4.1 Problem statement

In this section interconnected systems are studied that can be decomposed into subsystems Σ_i that belong to one of the following two sets:

- For the subsystems Σ_i, $i \in \mathcal{N}_o \subset \mathcal{N}$, the state $\boldsymbol{x}_i(t)$ is measurable.
- For the subsystems Σ_i, $i \in \mathcal{N}_d = \mathcal{N} \setminus \mathcal{N}_o$, the state $\boldsymbol{x}_i(t)$ is not accessible for measurement.

Hence, $\mathcal{N}_o$ and $\mathcal{N}_d$ are disjoint and $\mathcal{N}_o \cup \mathcal{N}_d = \mathcal{N}$. Regarding the subsystems Σ_i where the state $\boldsymbol{x}_i(t)$ is not measurable the following assumption is made:

A 5.1 For all Σ_i, $i \in \mathcal{N}_d$ the matrix $\boldsymbol{A}_i$ is Hurwitz and the initial condition $\boldsymbol{x}_i(0)$ is known to be bounded by

$$\|\boldsymbol{x}_i(0)\|_\infty \leq \bar{x}_{0i}. \tag{5.25}$$

For ease of exposition consider that the overall plant is transformed such that $\mathcal{N}_o = \{1, \ldots, N_o\}$ and $\mathcal{N}_d = \{N_d, \ldots, N\}$ with $N_d = N_o + 1$. Note that the transformation of the system in the proposed form is simply a reordering of the state and is, hence, always possible.

The aim of the event-based controller proposed in this section is formulated in terms of the performance of the event-based control system to be designed.

Definition 5.1 *The maximum deviation*

$$J := \sup_{t \geq 0} \|\boldsymbol{x}(t) - \boldsymbol{x}_{\mathrm{r}}(t)\|_{\infty} = \sup_{t \geq 0} \|\boldsymbol{e}(t)\|_{\infty}. \tag{5.26}$$

between the state $\boldsymbol{x}(t)$ of the plant (5.1a), (5.1b) *in the event-based control system to be designed and the state $\boldsymbol{x}_{\mathrm{r}}(t)$ of the reference system* (5.2) *is called performance J of the event-based control system.*

A requirement on the performance J of the event-based control system is expressed as

$$J \leq \bar{J}, \tag{5.27}$$

where $\bar{J} \in \mathbb{R}_{+}$ denotes the desired maximum deviation between the behavior of the event-based control system and the reference system (5.2).

The fact that only a part of the subsystem states is measurable has the following consequence for the structure of the event-based control system: Since the state $\boldsymbol{x}_i(t)$ of the subsystems Σ_i $(i \in \mathcal{N}_{\mathrm{d}})$ is not accessible, no event generators can be applied to these subsystems. This case is illustrated in Fig. 5.9 for Σ_N where $N \in \mathcal{N}_{\mathrm{d}}$ is considered. The figure shows that the constraints on the measurability of the subsystem states implies some restrictions on the control structure. The central question to be answered in this section concerns the not measurable states:

> Can the deviation of the not measurable states $\boldsymbol{x}_i(t)$ $(i \in \mathcal{N}_{\mathrm{d}})$ from the corresponding states $\boldsymbol{x}_{\mathrm{r}i}(t)$ of the reference system (5.2) be bounded by an event-based controller that has access to the measurable subsystem states only, such that the event-based control system satisfies the requirement (5.27) for a given bound $\bar{J}$?

Basic idea. In Sec. 5.4.4 it is shown that the maximum deviation between the not measurable states $\boldsymbol{x}_i(t)$ $(i \in \mathcal{N}_{\mathrm{d}})$ and the corresponding states of the reference system (5.2) depends upon the event thresholds $\bar{e}_j$ $(j \in \mathcal{N}_{\mathrm{o}})$. Hence, the main idea is that a desired control performance of the event-based control system can be achieved by appropriately adjusting the triggering conditions in the event generators E_i $(i \in \mathcal{N}_{\mathrm{o}})$. In other words, the loss of information about the states $\boldsymbol{x}_i(t)$ for all $i \in \mathcal{N}_{\mathrm{d}}$ shall be compensated by refining the accessible information $\boldsymbol{x}_j(t)$ from Σ_j for all $j \in \mathcal{N}_{\mathrm{o}}$.

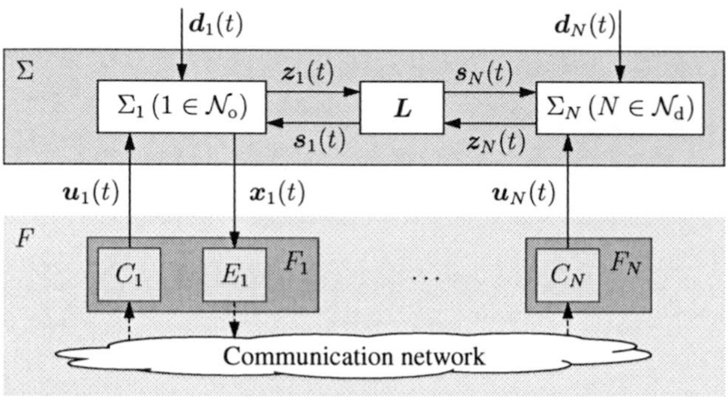

Figure 5.9: Structure of the event-based control loop where the state of only some subsystems is measurable

5.4.2 Structure of the event-based control system

The structure of the event-based control system is illustrated in Fig. 5.9. Note that this structure differs from the generalized one depicted in Fig. 4.1 in that the state $\boldsymbol{x}_i(t)$ of some subsystems Σ_i $(i \in \mathcal{N}_o)$ is not accessible (as for the example of Σ_N). According to the separation of the subsystems Σ_i into the two sets $\mathcal{N}_o$ and $\mathcal{N}_d$, the local controllers F_i of the respective subsystems have different structures:

- For the subsystems Σ_i, $i \in \mathcal{N}_o$, the local controller F_i consists of the control input generator C_i and the event generator E_i.
- For the subsystems Σ_i, $i \in \mathcal{N}_d$, where the subsystem state $\boldsymbol{x}_i(t)$ is not measurable the controller F_i only consists of the control input generator C_i.

The design of the control input generators and the event generators follows the idea of the method presented in Sec. 5.2.2 and is summarized in the following.

Control input generator C_i**.** The control input generator C_i (for all $i \in \mathcal{N}$) uses the model (5.3) of the overall reference system (5.2) to determine a prediction $\boldsymbol{x}_s(t)$ of the plant state $\boldsymbol{x}(t)$ for the time $t \in [t_k, t_{k+1})$. The control input generator C_i is represented by (5.4).

Event generator E_i**.** The event generator E_i (for $i \in \mathcal{N}_o$) also includes the model (5.3) and it continuously determines the difference

$$\boldsymbol{x}_{\Delta i}(t) = \boldsymbol{x}_i(t) - \boldsymbol{x}_{si}(t)$$

and triggers an event whenever the condition

$$\|\boldsymbol{x}_{\Delta i}(t)\|_\infty = \|\boldsymbol{x}_i(t) - \boldsymbol{x}_{\mathrm{s}i}(t)\|_\infty = \bar{e}_i, \quad i \in \mathcal{N}_\mathrm{o} \tag{5.28}$$

holds, where $\bar{e}_i \in \mathrm{I\!R}_+$ denotes the event threshold. The event generator E_i is described by the model (5.6).

Due to the event generation and the reset of the model state $\boldsymbol{x}_{\mathrm{s}i}(t)$, the difference state $\boldsymbol{x}_{\Delta i}(t)$ is bounded from above by

$$\sup_{t \geq 0} \|\boldsymbol{x}_{\Delta i}(t)\|_\infty = \bar{e}_i, \quad \forall\, i \in \mathcal{N}_\mathrm{o}. \tag{5.29}$$

Note that the difference $\boldsymbol{x}_{\Delta i}(t)$ for all $i \in \mathcal{N}_\mathrm{d}$ cannot be monitored and is, hence, not bounded by the event triggering and the state reset.

5.4.3 Performance of the event-based control system

The analysis presented in this section derives a bound on the performance J of the event-based control system (5.1a), (5.1b), (5.4), (5.6) in terms of the difference state $\boldsymbol{x}_\Delta(t)$.

This deviation $\boldsymbol{e}(t) = \boldsymbol{x}(t) - \boldsymbol{x}_\mathrm{s}(t)$ is described by the state-space model (5.17) which yields

$$\boldsymbol{e}(t) = \int_0^t \mathrm{e}^{\bar{\boldsymbol{A}}(t-\tau)} \boldsymbol{B}\boldsymbol{K}\boldsymbol{x}_\Delta(\tau)\mathrm{d}\tau.$$

From the last equation the relation

$$J = \sup_{t \geq 0} \|\boldsymbol{e}(t)\|_\infty \leq \kappa \cdot \sup_{t \geq 0} \|\boldsymbol{x}_\Delta(t)\|_\infty \tag{5.30}$$

follows with

$$\kappa = \int_0^\infty \left\| \mathrm{e}^{\bar{\boldsymbol{A}}\tau} \boldsymbol{B}\boldsymbol{K} \right\|_\infty \mathrm{d}\tau. \tag{5.31}$$

Since the matrix $\bar{\boldsymbol{A}}$ is Hurwitz by design, κ is finite. Inequality (5.30) shows that the performance J of the event-based system and the reference system depends upon the difference state $\boldsymbol{x}_\Delta(t)$. This result is summarized in the following lemma.

Lemma 5.1 *The maximum deviation J between the state $\boldsymbol{x}(t)$ of the event-based control system (5.1a), (5.1b), (5.4), (5.6) and the state $\boldsymbol{x}_\mathrm{r}(t)$ of the reference system (5.2) is bounded from above by (5.30), (5.31).*

With the result (5.30) the requirement on the performance (5.27) can be restated as the relation

$$\sup_{t\geq 0} \|\boldsymbol{x}_\Delta(t)\|_\infty \leq \frac{\bar{J}}{\kappa} \tag{5.32}$$

which is a requirement on the boundedness of the difference state $\boldsymbol{x}_\Delta(t)$. Note that a desired level of performance $\bar{J}$ can trivially be accomplished by means of the event triggering (5.28) and the reset of the model state $\boldsymbol{x}_\mathrm{s}(t)$ only if $\mathcal{N}_\mathrm{o} = \mathcal{N}$, because then Eq. (5.29) holds for all $i = 1, \ldots, N$ and the event thresholds $\bar{e}_i$ can be chosen such that the condition (5.32) is satisfied. However, since $\mathcal{N}_\mathrm{o} \subset \mathcal{N}$ is considered in this section, the difference states $\boldsymbol{x}_{\Delta i}(t)$ for all $i \in \mathcal{N}_\mathrm{d}$ are not bounded by (5.29), which raises the question how the performance requirement (5.27) can be satisfied in this case. This question will be answered in the next section.

5.4.4 Boundedness of the difference states

The previous section showed that a requirement on the performance J of the event-based control system (5.1a), (5.1b), (5.4), (5.6) can be expressed as a condition on the boundedness of the difference state $\boldsymbol{x}_\Delta(t)$. The analysis presented in this section derives a bound on the difference state $\boldsymbol{x}_{\Delta i}(t)$ $(i \in \mathcal{N}_\mathrm{d})$, which will be shown to be a function of the event thresholds $\bar{e}_j$ $(j \in \mathcal{N}_\mathrm{o})$. This result is used later to develop a design method for the event thresholds $\bar{e}_j$ that ensures a desired control performance $\bar{J}$.

First, consider the model Σ_s described in (5.3). The i-th component $\boldsymbol{x}_{\mathrm{s}i}(t)$ of the model state $\boldsymbol{x}_\mathrm{s}(t)$ for some $i \in \mathcal{N}_\mathrm{d}$ is described by the state-space model

$$\dot{\boldsymbol{x}}_{\mathrm{s}i}(t) = \boldsymbol{A}_i \boldsymbol{x}_{\mathrm{s}i}(t) - \boldsymbol{B}_i \boldsymbol{K}_i \boldsymbol{x}_\mathrm{s}(t) + \boldsymbol{E}_i \hat{\boldsymbol{d}}_{ik} + \boldsymbol{E}_{\mathrm{s}i} \boldsymbol{s}_{\mathrm{s}i}(t), \quad \boldsymbol{x}_{\mathrm{s}i}(0) = \boldsymbol{0} \tag{5.33a}$$

$$\boldsymbol{z}_{\mathrm{s}i}(t) = \boldsymbol{C}_{\mathrm{z}i} \boldsymbol{x}_{\mathrm{s}i}(t), \tag{5.33b}$$

for all $t \geq 0$ where the signal $\boldsymbol{s}_{\mathrm{s}i}(t)$ is given in Eq. (5.22). Due to the fact that the state $\boldsymbol{x}_i(t)$ of the system Σ_i cannot be measured, the initial state $\boldsymbol{x}_{\mathrm{s}i}(0)$ in the model (5.33) cannot be set to the current subsystem state $\boldsymbol{x}_i(0)$ and is, thus, initialized with zero. Moreover, note that the model state $\boldsymbol{x}_{\mathrm{s}i}(t)$ is not reset at the event times and, hence, the model (5.33) is valid for all $t \geq 0$. The subsystem Σ_i $(i \in \mathcal{N}_\mathrm{d})$ together with the associated control input generator C_i is represented by the state-space model

$$\dot{\boldsymbol{x}}_i(t) = \boldsymbol{A}_i \boldsymbol{x}_i(t) - \boldsymbol{B}_i \boldsymbol{K}_i \boldsymbol{x}_\mathrm{s}(t) + \boldsymbol{E}_i \boldsymbol{d}_i(t) + \boldsymbol{E}_{\mathrm{s}i} \boldsymbol{s}_i(t) \tag{5.34}$$

where the initial state $\boldsymbol{x}_i(0)$ satisfies $\|\boldsymbol{x}_i(0)\|_\infty \leq \bar{x}_{0i}$.

From (5.33) the model

$$\dot{\boldsymbol{x}}_{\Delta i}(t) = \boldsymbol{A}_i \boldsymbol{x}_{\Delta i}(t) + \boldsymbol{E}_i\big(\boldsymbol{d}_i(t) - \hat{\boldsymbol{d}}_{ik}\big) + \boldsymbol{E}_{\mathrm{s}i}\boldsymbol{s}_{\Delta i}(t) \tag{5.35}$$

$$\boldsymbol{s}_{\Delta i}(t) = \sum_{j=1}^{N} \boldsymbol{L}_{ij}\boldsymbol{C}_{\mathrm{z}j}\boldsymbol{x}_{\Delta j}(t) \tag{5.36}$$

for the difference state $\boldsymbol{x}_{\Delta i}(t)$ $(i \in \mathcal{N}_{\mathrm{d}})$ follows where for the initial state the relation (5.25) holds. With

$$\gamma_{0i} := \sup_{t \geq 0} \left\| \mathrm{e}^{\boldsymbol{A}_i t} \right\|_\infty \tag{5.37a}$$

$$\gamma_{\mathrm{d}i} := \int_0^\infty \left\| \mathrm{e}^{\boldsymbol{A}_i \tau} \boldsymbol{E}_i \right\|_\infty \mathrm{d}\tau \tag{5.37b}$$

$$\gamma_{ij} := \int_0^\infty \left\| \mathrm{e}^{\boldsymbol{A}_i \tau} \boldsymbol{E}_{\mathrm{s}i}\boldsymbol{L}_{ij}\boldsymbol{C}_{\mathrm{z}j} \right\|_\infty \mathrm{d}\tau. \tag{5.37c}$$

Eqs. (5.25), (5.35) yield

$$\sup_{t \geq 0} \|\boldsymbol{x}_{\Delta i}(t)\|_\infty \leq \gamma_{0i} \cdot \bar{x}_{0i} + \gamma_{\mathrm{d}i} \cdot \bar{d}_{\Delta i} + \sum_{j=1}^{N} \gamma_{ij} \cdot \sup_{t \geq 0} \|\boldsymbol{x}_{\Delta j}(t)\|_\infty, \quad \forall\, i \in \mathcal{N}_{\mathrm{d}}$$

where

$$\bar{d}_{\Delta i} \geq \left\| \boldsymbol{d}_i(t) - \hat{\boldsymbol{d}}_{ik} \right\|_\infty, \quad \forall\, t \geq 0, k \in \mathbb{N}_0$$

denotes the maximum deviation between the local disturbance $\boldsymbol{d}_i(t)$ and its estimate $\hat{\boldsymbol{d}}_{ik}$. To make this statement more precise, substitute (5.29) into the last inequality which results in

$$\sup_{t \geq 0} \|\boldsymbol{x}_{\Delta i}(t)\|_\infty \leq \gamma_{0i} \cdot \bar{x}_{0i} + \gamma_{\mathrm{d}i} \cdot \bar{d}_{\Delta i} + \sum_{j \in \mathcal{N}_{\mathrm{o}}} \gamma_{ij} \cdot \bar{e}_j + \sum_{j \in \mathcal{N}_{\mathrm{d}} \setminus \{i\}} \gamma_{ij} \cdot \sup_{t \geq 0} \|\boldsymbol{x}_{\Delta j}(t)\|_\infty . \tag{5.38}$$

Note that (5.38) is an implicit expression for the bound on $\boldsymbol{x}_{\Delta i}(t)$ for all $i \in \mathcal{N}_{\mathrm{d}}$ because it is a function of the bounds on $\boldsymbol{x}_{\Delta j}(t)$ for all $j \in \mathcal{N}_{\mathrm{d}} \setminus \{i\}$. In the following an explicit formulation of the bounds on the difference states $\boldsymbol{x}_{\Delta i}(t)$ for all $i \in \mathcal{N}_{\mathrm{d}}$ is derived. With

$$\delta_i := \sup_{t \geq 0} \|\boldsymbol{x}_{\Delta i}(t)\|_\infty, \quad i \in \mathcal{N}_{\mathrm{d}} \tag{5.39}$$

Eq. (5.38) yields

$$\boldsymbol{\Psi}_{\mathrm{d}} \begin{pmatrix} \delta_{N_\mathrm{d}} \\ \vdots \\ \delta_N \end{pmatrix} \leq \begin{pmatrix} \gamma_{0N_\mathrm{d}} \cdot \bar{x}_{0N_\mathrm{d}} + \gamma_{\mathrm{d}N_\mathrm{d}} \cdot \bar{d}_{\Delta N_\mathrm{d}} + \sum_{j\in\mathcal{N}_\mathrm{o}} \gamma_{N_\mathrm{d}j} \cdot \bar{e}_j \\ \vdots \\ \gamma_{0N} \cdot \bar{x}_{0N} + \gamma_{\mathrm{d}N} \cdot \bar{d}_{\Delta N} + \sum_{j\in\mathcal{N}_\mathrm{o}} \gamma_{Nj} \cdot \bar{e}_j \end{pmatrix}, \tag{5.40}$$

with the matrix

$$\boldsymbol{\Psi}_{\mathrm{d}} = \begin{pmatrix} 1 & \dots & -\gamma_{N_\mathrm{d}N} \\ \vdots & \ddots & \vdots \\ -\gamma_{NN_\mathrm{d}} & \dots & 1 \end{pmatrix}. \tag{5.41}$$

Under the condition that $\boldsymbol{\Psi}_{\mathrm{d}}$ is an M-matrix, the inverse $\boldsymbol{\Psi}_{\mathrm{d}}^{-1}$ exists and is non-negative and, thus, the relation (5.40) can be reformulated as

$$\begin{pmatrix} \delta_{N_\mathrm{d}} \\ \vdots \\ \delta_N \end{pmatrix} \leq \boldsymbol{\Psi}_{\mathrm{d}}^{-1} \begin{pmatrix} \gamma_{0N_\mathrm{d}} \cdot \bar{x}_{0N_\mathrm{d}} + \gamma_{\mathrm{d}N_\mathrm{d}} \cdot \bar{d}_{\Delta N_\mathrm{d}} + \sum_{j\in\mathcal{N}_\mathrm{o}} \gamma_{N_\mathrm{d}j} \cdot \bar{e}_j \\ \vdots \\ \gamma_{0N} \cdot \bar{x}_{0N} + \gamma_{\mathrm{d}N} \cdot \bar{d}_{\Delta N} + \sum_{j\in\mathcal{N}_\mathrm{o}} \gamma_{Nj} \cdot \bar{e}_j \end{pmatrix}. \tag{5.42}$$

Inequality (5.42) shows that the upper bound on the difference state $\boldsymbol{x}_{\Delta i}(t)$ for all $i \in \mathcal{N}_\mathrm{d}$ can be manipulated by appropriately setting the event thresholds $\bar{e}_j$ $(j \in \mathcal{N}_\mathrm{o})$.

Lemma 5.2 *Consider the event-based control system system* (5.1a), (5.1b), (5.4), (5.6) *where the states $\boldsymbol{x}_i(t)$ of the subsystems Σ_i $(i \in \mathcal{N}_\mathrm{o})$ are measurable and the states $\boldsymbol{x}_j(t)$ of the remaining subsystems Σ_j $(j \in \mathcal{N}_\mathrm{d})$ are not accessible for measurement. The difference states $\boldsymbol{x}_{\Delta i}(t)$ for all $i \in \mathcal{N}_\mathrm{d}$ in the event-based control system are bounded from above by $\|\boldsymbol{x}_{\Delta i}(t)\|_\infty \leq \delta_i$ with δ_i according to Eq.* (5.42).

5.4.5 Event threshold design method

The result summarized in Lemma 5.2 will now be used to develop a method for the design of the event thresholds $\bar{e}_j$ $(j \in \mathcal{N}_\mathrm{o})$, such that a desired performance (5.27) is guaranteed. In order to specify the aim of this design method, note that the requirement (5.32) on the difference state $\boldsymbol{x}_\Delta(t)$ can be expressed as a requirement on the difference states $\boldsymbol{x}_{\Delta i}(t)$ for all $i \in \mathcal{N}$:

$$\sup_{t\geq 0} \|\boldsymbol{x}_{\Delta i}(t)\|_\infty \leq \frac{\bar{J}}{\kappa}, \quad \text{for all } i = 1, \dots, N.$$

The last inequality together with (5.29), (5.39) implies that the design aim is to find event thresholds $\bar{e}_i$ that satisfy the relation

$$0 < \bar{e}_i \leq \frac{\bar{J}}{\kappa}, \qquad \text{for all } i \in \mathcal{N}_{\mathrm{o}} \tag{5.43}$$

and which guarantee that

$$0 \leq \delta_i \leq \frac{\bar{J}}{\kappa} \qquad \text{for all } i \in \mathcal{N}_{\mathrm{d}} \tag{5.44}$$

holds. The aim of the next investigation is to formulate the requirement (5.44) in dependence upon the event thresholds $\bar{e}_i$.

Let

$$\bar{\boldsymbol{e}} = \begin{pmatrix} \bar{e}_1 & \dots & \bar{e}_{N_{\mathrm{o}}} \end{pmatrix}^\top$$

denote the vector of variables to be determined. Then the relation (5.40) can be restated explicitly dependent on the vector $\bar{\boldsymbol{e}}$ as

$$\boldsymbol{\Psi}_{\mathrm{d}} \cdot \boldsymbol{\delta} \leq \boldsymbol{\psi} + \boldsymbol{\Psi}_{\mathrm{o}} \cdot \bar{\boldsymbol{e}} \tag{5.45}$$

with $\boldsymbol{\delta} = \begin{pmatrix} \delta_{N_{\mathrm{d}}} \dots \delta_N \end{pmatrix}^\top$, the matrix $\boldsymbol{\Psi}_{\mathrm{d}}$ as defined in (5.41) and

$$\boldsymbol{\psi} = \begin{pmatrix} \gamma_{0N_{\mathrm{d}}} \cdot \bar{x}_{0N_{\mathrm{d}}} + \gamma_{\mathrm{d}N_{\mathrm{d}}} \cdot \bar{d}_{\Delta N_{\mathrm{d}}} \\ \vdots \\ \gamma_{0N} \cdot \bar{x}_{0N} + \gamma_{\mathrm{d}N} \cdot \bar{d}_{\Delta N} \end{pmatrix}, \qquad \boldsymbol{\Psi}_{\mathrm{o}} = \begin{pmatrix} \gamma_{N_{\mathrm{d}}1} & \dots & \gamma_{N_{\mathrm{d}}N_{\mathrm{o}}} \\ \vdots & \ddots & \vdots \\ \gamma_{N1} & \dots & \gamma_{NN_{\mathrm{o}}} \end{pmatrix}. \tag{5.46}$$

Note that the requirement (5.44) can be reformulated as

$$\boldsymbol{\delta} \leq \mathbf{1} \cdot \frac{\bar{J}}{\kappa}.$$

Consequently, the relation (5.45) implies that the requirement (5.44) is fulfilled if the event thresholds in $\bar{\boldsymbol{e}}$ are chosen such that the inequality

$$\boldsymbol{\psi} + \boldsymbol{\Psi}_{\mathrm{o}} \cdot \bar{\boldsymbol{e}} \leq \boldsymbol{\Psi}_{\mathrm{d}} \cdot \mathbf{1} \cdot \frac{\bar{J}}{\kappa} \tag{5.47}$$

is satisfied.

Theorem 5.3 *If the vector $\bar{e}$ of event thresholds $\bar{e}_i$ $(i \in \mathcal{N}_\text{o})$ fulfills the inequalities (5.43), (5.47), then the event-based control system (5.1a), (5.1b) with control input generators C_i as in (5.4) for all Σ_i $(i \in \mathcal{N})$ and event generators E_i given in (5.6) for all Σ_i $(i \in \mathcal{N}_\text{o})$ satisfies the desired level of performance (5.27).*

In addition to the conditions (5.43), (5.47) it is often also desirable to maximize the event thresholds $\bar{e}_i$ in order to obtain larger minimum inter-event times as described in Theorem 5.2 in Section 5.2.6. These constraints are now used in order to formulate the event threshold design method as the linear programming problem [112]

$$\max \quad \boldsymbol{c}^\top \bar{\boldsymbol{e}} \tag{5.48a}$$

$$\text{s.t.} \quad \boldsymbol{\Psi}_\text{o} \cdot \bar{\boldsymbol{e}} \leq \boldsymbol{\Psi}_\text{d} \cdot \mathbf{1} \cdot \frac{\bar{J}}{\kappa} - \boldsymbol{\psi}, \tag{5.48b}$$

$$\bar{\boldsymbol{e}} \leq \mathbf{1} \cdot \frac{\bar{J}}{\kappa}, \tag{5.48c}$$

$$\bar{\boldsymbol{e}} > 0 \tag{5.48d}$$

where $\boldsymbol{c} \in \mathbb{R}_+^{N_\text{o}}$ is a vector of weighting factors.

Feasibility of the linear programming problem. In due consideration of the constraints (5.48b)–(5.48d), the problem (5.48) is feasible only if

- both the bound $\bar{x}_{0i}$ on the initial condition and the bound $\bar{d}_{\Delta i}$ on the maximum disturbance estimation error for all $i \in \mathcal{N}_\text{d}$ are small with respect to the desired performance level $\bar{J}$ and

- the subsystems Σ_i and Σ_j for all $i, j \in \mathcal{N}_\text{d}$ are weakly coupled in a sense defined below.

These requirements are necessary conditions for the feasibility of the problem (5.48) which are now explained in detail. Consider the constraint (5.48b) and observe that the right-hand side of the inequality is required to be element-wise positive which implies the requirement

$$\underbrace{\begin{pmatrix} 1 & \dots & -\gamma_{N_\text{d}N} \\ \vdots & \ddots & \vdots \\ -\gamma_{NN_\text{d}} & \dots & 1 \end{pmatrix}}_{\boldsymbol{\Psi}_\text{d}} \underbrace{\begin{pmatrix} \frac{\bar{J}}{\kappa} \\ \vdots \\ \frac{\bar{J}}{\kappa} \end{pmatrix}}_{\mathbf{1} \cdot \frac{\bar{J}}{\kappa}} > \underbrace{\begin{pmatrix} \gamma_{0N_\text{d}} \cdot \bar{x}_{0N_\text{d}} + \gamma_{\text{d}N_\text{d}} \cdot \bar{d}_{\Delta N_\text{d}} \\ \vdots \\ \gamma_{0N} \cdot \bar{x}_{0N} + \gamma_{\text{d}N} \cdot \bar{d}_{\Delta N} \end{pmatrix}}_{\boldsymbol{\psi}}. \tag{5.49}$$

This inequality can be satisfied only if the left-hand side is element-wise positive (since ψ is element-wise positive) which results in the condition

$$\sum_{j \in \mathcal{N}_\mathrm{d}} \gamma_{ij} < 1, \quad \text{for all } i \in \mathcal{N}_\mathrm{d}. \tag{5.50}$$

This inequality can be interpreted as a demand for weak coupling between the subsystems Σ_i and Σ_j for all $i, j \in \mathcal{N}_\mathrm{d}$. Assume that the condition (5.50) is fulfilled, then it can be immediately concluded from (5.49) that for a given performance requirement $\bar{J}$ the bounds $\bar{x}_{0i}$ and $\bar{d}_{\Delta i}$ for all $i \in \mathcal{N}_\mathrm{d}$ must be sufficiently small.

Design algorithm. The method for the design of the event thresholds $\bar{e}_i$ for $i \in \mathcal{N}_\mathrm{d}$ is now summarized in the following algorithm.

Algorithm 5.1 (Design of the event thresholds $\bar{e}_i$ $(i \in \mathcal{N}_\mathrm{o})$)

Given: Interconnected system (5.1a), (5.1b),
bounds $\bar{x}_{i0}$ on the initial states $\boldsymbol{x}_i(0)$ for all $i \in \mathcal{N}_\mathrm{d}$,
bounds $\bar{d}_{\Delta i}$ on the maximum disturbance estimation error for all $i \in \mathcal{N}$ and
performance requirement $\bar{J}$.

1. Check whether the subsystems Σ_i $(i \in \mathcal{N}_\mathrm{d})$ are stable. If one or more subsystems Σ_i $(i \in \mathcal{N}_\mathrm{d})$ are unstable, stop (the performance requirement cannot be satisfied by any choice of event thresholds $\bar{e}_i$ for $i \in \mathcal{N}_\mathrm{o}$).
2. Determine κ as in (5.31) and $\gamma_{0i}, \gamma_{\mathrm{d}i}$ and γ_{ij} as in (5.37). Construct the matrices $\boldsymbol{\Psi}_\mathrm{d}, \boldsymbol{\Psi}_\mathrm{o}$ and the vector $\boldsymbol{\psi}$ according to (5.41) and (5.46).
3. Check whether the conditions (5.49), (5.50) are satisfied. If one of these conditions does not hold, stop (the design method is not applicable).
4. Choose a vector $\boldsymbol{c} \in \mathbb{R}_+^{N_\mathrm{o}}$ of weighting factors and solve the linear programming problem (5.48).

Result: Event thresholds $\bar{e}_i$ $(i \in \mathcal{N}_\mathrm{o})$ which satisfy the performance requirement (5.27) for the event-based control system (5.1a), (5.1b) with control input generators C_i as in (5.4) for all Σ_i $(i \in \mathcal{N})$ and event generators E_i given in (5.6) for all Σ_i $(i \in \mathcal{N}_\mathrm{o})$.

The following summarizes some comments and remarks on the proposed event threshold design method:

- In general the obtained solution $\bar{e}$ for the problem (5.48) is not unique. The result $\bar{e}$ of the design algorithm can be adapted by choosing a different vector c of weighting factors and repeating step 4.
- If the algorithm stops in step 3, because the conditions (5.49), (5.50) are not satisfied, it does not mean that there does not exist a set of event thresholds $\bar{e}_i$ $(i \in \mathcal{N}_o)$ which solves the problem (5.27), but a solution cannot be found with the proposed method.
- The event threshold design method can be extended in order to guarantee certain bounds on the minimum inter-event times $T_{\min i}$ $(i \in \mathcal{N}_o)$ which, from the analysis in Sec. 5.2.6, are known to depend upon the bounds on the difference states $x_{\Delta i}(t)$ for all $i \in \mathcal{N}$. To this end, the constraint (5.48d) must be replaced by

$$\bar{e} \geq \bar{e}_{\min},$$

where the vector $\bar{e}_{\min}$ is chosen so as to ensure desired minimum inter-event times. However, this constraint further restricts the feasibility of the linear programming problem (5.48).

5.4.6 Application example: Interconnected two-tank system

The proposed event-threshold design method is now tested on the system depicted in Fig. 5.10 which is a simplified version of the process described in Sec. 2.4. The original process has been modified for the purpose of explaining the main characteristics of the design method.

Process description. The modified process consists of two reactors TB and TS, in each of which the level $l(t)$ and the temperature $\vartheta(t)$ of the liquid shall be controlled. The level $l_{\mathrm{B}}(t)$ and temperature $\vartheta_{\mathrm{B}}(t)$ of the content in TB are considered to be measurable, whereas the level $l_{\mathrm{S}}(t)$ and temperature $\vartheta_{\mathrm{S}}(t)$ in TS are not measurable. The reactor TB is fed by the water supply from T_1 from where the inflow can be controlled by means of the valve angle $u_{\mathrm{T1}}(t)$. Analogously, reactor TS is connected to the tank T_3 from where the flow to TS can be controlled by the valve angle $u_{\mathrm{T3}}(t)$. Heating rods in both reactors are applied to increase the temperature of the content by the inputs $u_{\mathrm{H}}(t)$ and $u_{\mathrm{S}}(t)$. The process is disturbed by an additional and undesired inflow into reactor TB from the supply FW. This disturbance can be interpreted as an inflow, caused by a blockage of the valve that connects FW with TB, which is assumed to be bounded by

$$|d(t)| \leq 0.05, \quad t \geq 0. \tag{5.51}$$

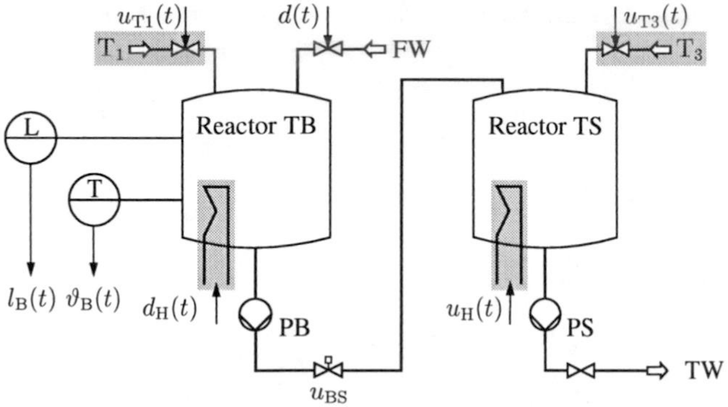

Figure 5.10: Simplified thermofluid process

For the following investigation, the overall system is decomposed into four scalar subsystems with the states

$$x_1(t) = l_B(t), \quad x_2(t) = \vartheta_B(t), \quad x_3(t) = l_S(t), \quad x_4(t) = \vartheta_S(t).$$

Each subsystem is described by a linear state-space model (5.1a) with

$$\begin{aligned} \boldsymbol{A}_1 &= -0.003, & \boldsymbol{B}_1 &= 0.14, & \boldsymbol{E}_1 &= 0.26, \\ \boldsymbol{A}_2 &= -0.006, & \boldsymbol{B}_2 &= 0.03, & \boldsymbol{E}_2 &= -0.11, \\ \boldsymbol{A}_3 &= -0.005, & \boldsymbol{B}_3 &= 0.13, & \boldsymbol{E}_3 &= 0, & \quad (5.52a) \\ \boldsymbol{A}_4 &= -0.006, & \boldsymbol{B}_4 &= 0.03, & \boldsymbol{E}_4 &= 0 & \quad (5.52b) \end{aligned}$$

and $\boldsymbol{E}_{si} = \boldsymbol{C}_{zi} = 1$ for all $i = 1, \ldots, 4$. The subsystems are interconnected according to the relation (5.1b) with

$$\boldsymbol{L} = 10^{-3} \begin{pmatrix} 0 & 0 & 0 & 0 \\ -3.76 & 0 & 0 & 0 \\ 5.62 & 0 & 0 & 0 \\ 0 & 4.71 & -4.16 & 0 \end{pmatrix}.$$

Note that the states $x_1(t)$ and $x_2(t)$ of Σ_1 and Σ_2, respectively, are measurable, while the states $x_3(t)$ and $x_4(t)$ of Σ_3 and Σ_4 are unknown. Hence $\mathcal{N}_o = \{1, 2\}$ and $\mathcal{N}_d = \{3, 4\}$. For the following investigation it is assumed that for the initial states $x_3(0) = 0$ and $x_4(0) = 0$ holds and that $x_3(0)$ and $x_4(0)$ are known without uncertainty: $\bar{x}_3 = \bar{x}_4 = 0$.

Remark 5.1 *In this example the information that is available from the overall system is restricted to*

$$\begin{pmatrix} x_1(t) \\ x_2(t) \end{pmatrix} = \underbrace{\begin{pmatrix} 1 & 0 & 0 & 0 \\ 0 & 1 & 0 & 0 \end{pmatrix}}_{=:\, \boldsymbol{C}} \boldsymbol{x}(t).$$

Note that the pair $(\boldsymbol{A}, \boldsymbol{C})$ *with the matrix*

$$\boldsymbol{A} = \mathrm{diag}\,(\boldsymbol{A}_1, \ldots, \boldsymbol{A}_4) + \mathrm{diag}\,(\boldsymbol{E}_{\mathrm{s}1}, \ldots, \boldsymbol{E}_{\mathrm{s}4})\ \boldsymbol{L}\,\mathrm{diag}\,(\boldsymbol{C}_{\mathrm{z}1}, \ldots, \boldsymbol{C}_{\mathrm{z}4})$$

is not observable and, thus, the overall system state $\boldsymbol{x}(t)$ *cannot be reconstructed by an observer. This implies that the observer-based approaches to event-based output-feedback control [101, 146, 150] are not applicable to the system considered in this example.*

Controller and event threshold design. The following introduces a continuous reference system. Assume only for the purpose of the design of this system that all states are measurable. The interconnected subsystems (5.1a), (5.1b) with the parameters given in the previous paragraph controlled by a the state feedback with feedback gain

$$\boldsymbol{K} = \mathrm{diag}\,(0.1,\ 0.8,\ 0.2,\ 0.7)$$

yields the reference system

$$\dot{\boldsymbol{x}}_{\mathrm{r}}(t) = \underbrace{10^{-3} \begin{pmatrix} -17 & 0 & 0 & 0 \\ -3.8 & -30 & 0 & 0 \\ 5.6 & 0 & -31 & 0 \\ 0 & 9.4 & -12.5 & -27 \end{pmatrix}}_{= \bar{\boldsymbol{A}}} \boldsymbol{x}_{\mathrm{r}} + \underbrace{\begin{pmatrix} 0.26 \\ -0.11 \\ 0 \\ 0 \end{pmatrix}}_{= \boldsymbol{E}} d(t), \qquad \boldsymbol{x}_{\mathrm{r}}(0) = \boldsymbol{x}_0.$$

This reference system is assumed to have a desired disturbance behavior. A model of this system is included in each control input generator C_i $(i \in \mathcal{N})$ and each event generator E_i $(i \in \mathcal{N}_{\mathrm{d}})$. For ease of exposition, the generators do not include a disturbance estimation. Thus, in the models (5.3) the trivial estimate $\hat{d}_k \equiv 0$ is applied for all $k \in \mathbb{N}_0$. From Eq. (5.51) the bound

$$\bar{d}_\Delta = \sup_{t \geq 0} \left| d(t) - \hat{d}_k \right| = 0.05$$

follows. The event-based control system shall satisfy the performance requirement (5.27) with the choice

$$\bar{J} = 1.5. \tag{5.53}$$

The following explains how event thresholds $\bar{e}_1$ and $\bar{e}_2$ can be found by means of the Algorithm 5.1.

Step 1: From (5.52) it can be seen that the subsystems Σ_3 and Σ_4 are open-loop stable.

Step 2: According to (5.31) $\kappa = 1.15$ is obtained and the determination of the matrices $\boldsymbol{\Psi}_{\rm d}$, $\boldsymbol{\Psi}_{\rm o}$ and the vector $\boldsymbol{\psi}$ yields

$$\boldsymbol{\Psi}_{\rm d} = \begin{pmatrix} 1 & 0 \\ -0.69 & 1 \end{pmatrix}, \quad \boldsymbol{\Psi}_{\rm o} = \begin{pmatrix} 1.12 & 0 \\ 0 & 0.79 \end{pmatrix}, \quad \boldsymbol{\psi} = \begin{pmatrix} 0 \\ 0 \end{pmatrix}.$$

Step 3: Condition (5.50) is satisfied, since $\gamma_{34} = 0$ and $\gamma_{43} = 0.69 < 1$. In view of $\boldsymbol{\psi} = \mathbf{0}$, condition (5.49) is also fulfilled.

Step 4: With the vector $\boldsymbol{c} = \begin{pmatrix} 1 & 1 \end{pmatrix}^{\top}$ of weighting factors the linear programming problem (5.48) has the solution

$$\bar{e}_1 = 1.2, \quad \bar{e}_2 = 0.5. \tag{5.54}$$

for which with Eq. (5.42)

$$\delta_3 := \sup_{t \geq 0} \|\boldsymbol{x}_{\Delta 3}(t)\|_{\infty} \leq 1.3 = \frac{\bar{J}}{\kappa} \tag{5.55}$$

$$\delta_4 := \sup_{t \geq 0} \|\boldsymbol{x}_{\Delta 4}(t)\|_{\infty} \leq 1.3 = \frac{\bar{J}}{\kappa} \tag{5.56}$$

is obtained.

Simulation results. In this section the result (5.54) obtained by the proposed event threshold design method is evaluated by means of a simulation. The following analysis investigates the behavior of the event-based control loop subject to a constant disturbance $d(t) = 0.05$. The simulation results are illustrated in Fig. 5.11. The figures on the left-hand side show the level $x_1(t)$, the temperature $x_2(t)$ in reactor TB and the events triggered by the event generator of Σ_1 (e_l) and of Σ_2 (e_ϑ). The figures on the right-hand side depict the level $x_3(t)$ and temperature $x_4(t)$ in reactor TS. In each of these figures the plant state $\boldsymbol{x}(t)$ is represented by the solid line and the model state $\boldsymbol{x}_{\rm s}(t)$ by the dashed line. The disturbance $d(t)$ directly affects subsystems Σ_1 and Σ_2 which is reflected in the deviation between $x_1(t)$ and $x_{\rm s1}(t)$ or $x_2(t)$ and $x_{\rm s2}(t)$, re-

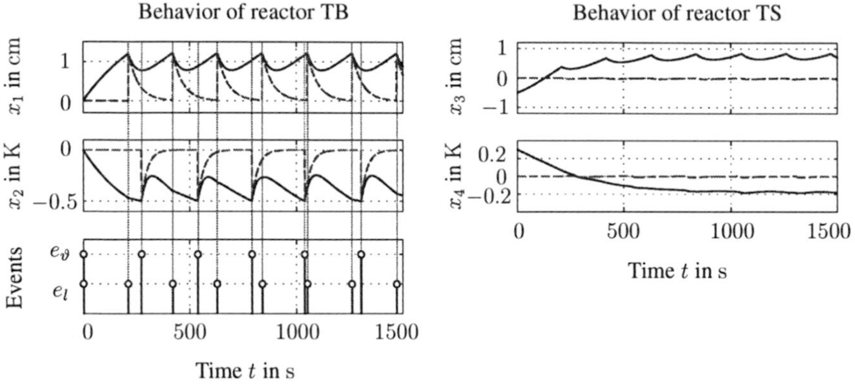

Figure 5.11: Behavior of the event-based control system where only the states $x_1(t)$ and $x_2(t)$ are measurable subject to a constant disturbance $d(t) = 0.05$

spectively. An event is triggered in either of the subsystems whenever the deviation state $x_{\Delta 1}(t)$ or $x_{\Delta 2}(t)$ reaches the respective event threshold specified in (5.54).

The trajectory of the deviation state $\boldsymbol{x}_\Delta(t)$ for all subsystems is illustrated in Fig. 5.12. The left-hand side of Fig. 5.12 shows the deviation states $x_{\Delta 1}(t)$ and $x_{\Delta 2}(t)$ for Σ_1 and Σ_2, respectively. Each time one of the deviation states $x_{\Delta 1}(t)$ or $x_{\Delta 2}(t)$ attains a bound of the rectangle denoting the event thresholds (5.54), the respective state is reset to zero. For each subsystem the deviation state is obviously bounded due to the event generation and reinitialization. In contrast to this, the right-hand side of Fig. 5.12 shows the trajectory of the deviation states $x_{\Delta 3}(t)$ and $x_{\Delta 4}(t)$ for the subsystems Σ_3 and Σ_4, respectively, where none of the deviation states attains the

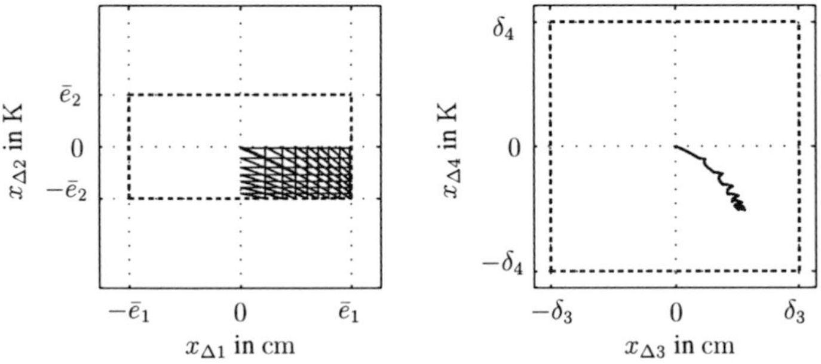

Figure 5.12: Deviation state $\boldsymbol{x}_\Delta(t)$

rectangle that marks the bounds (5.55). This implies that the performance requirement (5.27) with $\bar{J} = 1.5$ is satisfied by the derived event thresholds (5.54).

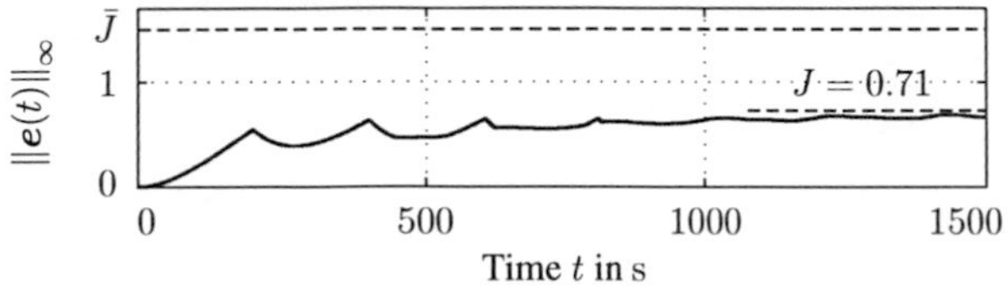

Figure 5.13: Deviation between the event-based control system and the reference system.

Figure 5.13 shows the deviation $\|e(t)\|_\infty$ between behavior of the event-based control system with the determined event thresholds (5.54) and the reference system (5.2). The maximum deviation between the behavior of both systems in the investigated time interval is $J = 0.71$, thus, considerably smaller than the performance requirement $\bar{J} = 1.5$. This example shows that the proposed event threshold design method might yield conservative results. This conservatism is attributable to the fact that the design method is developed from a worst-case analysis of the behavior of the event-based control system.

6 Event-based state-feedback control with local information couplings

This chapter investigates a decentralized event-based state-feedback approach from two different viewpoints. First, it proposes a method for the implementation of a given decentralized state-feedback law for the overall control system in an event-based manner, where coupling input estimation is used to reduce the number of events. Second, it investigates the stability of physically interconnected event-based state-feedback loops that are designed decentrally based on the assumption of vanishing interconnections. A small-gain condition is derived that yields an upper bound on the strength of the interconnections for which the stability of the isolated event-based control loops imply the stability of the overall control system. The chapter closes with the presentation and the discussion of an experiment that shows the behavior of the decentralized event-based state feedback applied to the thermofluid process.

6.1 Event-based control with unicast communication

This chapter investigates the event-based control of physically interconnected systems where the communication network is used to transmit information locally only, i.e., from an event generator E_i to the corresponding control input generator C_i but to no other component. This communication scheme illustrated in Fig. 6.1 for the example of a local information transmission from E_2 to C_2. Such a point-to-point information transmission is called unicast communication [151]. There are several situations where unicasting, as opposed to broadcasting, might be preferable from both a communication and a control perspective:

- If the communication network is deployed with a protocol that does not inherently provide broadcasting, the simultaneous transmission of information to all nodes of the network is costly with respect to the communication effort. In that case, the aim of the design of the communication topology is to reduce the overall number of communication links

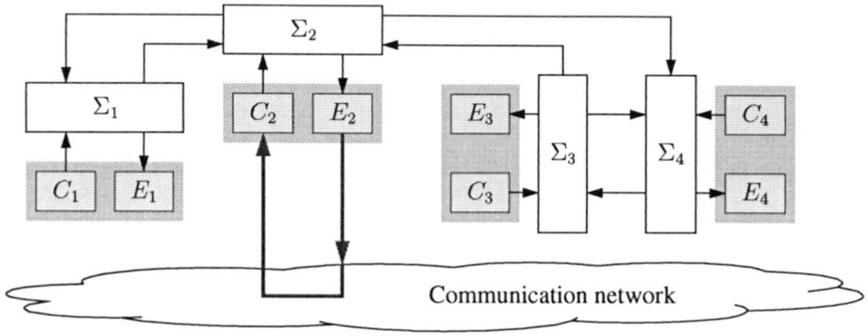

Figure 6.1: Unicast communication scheme

and, thus, unicasting represents the most desirable option. Another reason why unicasting might be preferred is the security aspect. In order to protect the communication between transmitter and receiver from attacks, a *virtual private network* (VPN) is often used which emulates a point-to-point connection [151].

- Consider a system that is composed of physically interconnected subsystems where subsystems can be added or removed from the overall system. The highest flexibility with respect to a varying system architecture is obtained if the controller for each subsystem is designed using model information of the very subsystem only and if the feedback communication occurs locally.

Section 6.2 proposes a method for the implementation of a given decentralized state-feedback law in an event-based manner. For the design of the control units, only local model information is used which implies that the triggering of events by the event generator E_i is not only caused by the disturbance $\boldsymbol{d}_i(t)$ but also by the coupling input $\boldsymbol{s}_i(t)$. Section 6.2 focuses on the effect of the coupling input $\boldsymbol{s}_i(t)$ and assumes the impact of the disturbance $\boldsymbol{d}_i(t)$ to be neglectable. Two methods for the coupling input estimation are proposed which aim at the further reduction of the feedback communication effort. Both methods are evaluated in a simulation for the example of the thermofluid process.

While Sec. 6.2 focuses on the event-based implementation of a given stabilizing decentralized state-feedback law, Sec. 6.3 investigates the behavior of interconnected event-based state-feedback loops that are designed decentrally assuming that the coupling input $\boldsymbol{s}_i(t)$ is absent. The main analysis problem concerns the question, on what condition on the interconnection between the subsystems the stability of the isolated event-based state-feedback loops implies the stability of the overall event-based control system. As the main analysis result, a condition is presented that states an upper bound on the interconnection matrix $\boldsymbol{L}$ for which the stability of the overall control system is guaranteed. Moreover, methods for the analysis of the boundedness of the plant state $x(t)$ and the minimum inter-event times are given.

Section 6.4 presents the result of an experiment that investigates the disturbance behavior of the decentralized event-based state feedback applied to the thermofluid process. It proves the analysis results of Sec. 6.3 to be marginally conservative in the presence of model uncertainties.

6.2 Decentralized event-based control

In this section event-based control of interconnected systems is investigated where the communication network is used to transmit information locally, as opposed to the broadcast communication structure studied in Ch. 5. In this context *local information transmission* means that the event generator E_i of subsystem Σ_i transmits the current subsystem state $\boldsymbol{x}_i(t_{k_i})$ at the event time t_{k_i} only to the corresponding control input generator C_i but to no other component of the overall event-based controller F (Fig. 6.2). Since the local control unit F_i generates the control input $\boldsymbol{u}_i(t)$ using local information only, the resulting overall controller F is called *decentralized event-based state-feedback controller.*

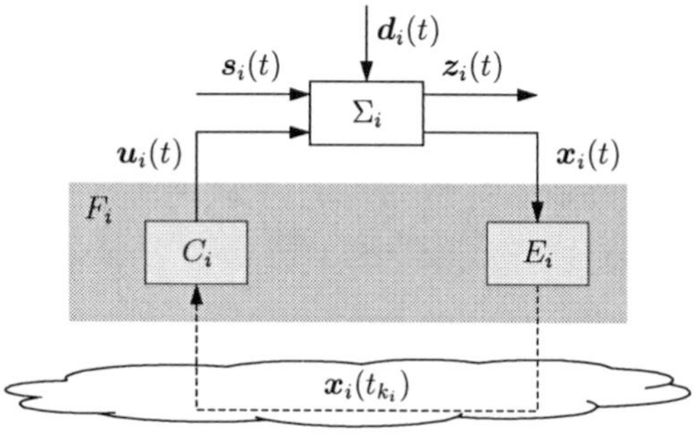

Figure 6.2: Local control unit F_i of the decentralized event-based state-feedback controller

6.2.1 Decentralized reference system

Consider the overall plant (2.1) together with the decentralized state feedback

$$\begin{pmatrix} \boldsymbol{u}_1(t) \\ \vdots \\ \boldsymbol{u}_N(t) \end{pmatrix} = \begin{pmatrix} \boldsymbol{K}_{\mathrm{d}1} & & \\ & \ddots & \\ & & \boldsymbol{K}_{\mathrm{d}N} \end{pmatrix} \begin{pmatrix} \boldsymbol{x}_1(t) \\ \vdots \\ \boldsymbol{x}_N(t) \end{pmatrix} \tag{6.1}$$

where $\boldsymbol{K}_{\mathrm{d}i} \in \mathbb{R}^{m_i \times n_i}$. The control law (6.1) is assumed to accomplish a desired performance of the closed-loop system. With the state-feedback gain

$$\boldsymbol{K}_{\mathrm{d}} := \boldsymbol{K} = \mathrm{diag}\left(\boldsymbol{K}_{\mathrm{d}1}, \ldots, \boldsymbol{K}_{\mathrm{d}N}\right) \tag{6.2}$$

the reference system (3.1) can be reformulated as

$$\Sigma_{\rm r}: \quad \dot{\boldsymbol{x}}_{\rm r}(t) = \underbrace{\left(\boldsymbol{A} - \boldsymbol{B}\boldsymbol{K}_{\rm d}\right)}_{:=\bar{\boldsymbol{A}}} \boldsymbol{x}_{\rm r}(t) + \boldsymbol{E}\boldsymbol{d}(t), \quad \boldsymbol{x}_{\rm r}(0) = \boldsymbol{x}_0, \tag{6.3}$$

where, as before, all signals are indicated with index r in order to distinguish them from the corresponding signals in the event-based control system that is investigated in the next section. In view of the decentralized state feedback (6.1), the system (6.3) is subsequently referred to as the *decentralized reference system.*

The system (6.3) can equivalently be stated in form of the interconnection-oriented model (2.3), (2.4) where the i-th subsystem is represented by the state-space model

$$\Sigma_{\mathrm{r}i}: \begin{cases} \dot{\boldsymbol{x}}_{\mathrm{r}i}(t) = \underbrace{\left(\boldsymbol{A}_i - \boldsymbol{B}_i\boldsymbol{K}_{\mathrm{d}i}\right)}_{:=\bar{\boldsymbol{A}}_i} \boldsymbol{x}_{\mathrm{r}i}(t) + \boldsymbol{E}_i\boldsymbol{d}_i(t) + \boldsymbol{E}_{\mathrm{s}i}\boldsymbol{s}_{\mathrm{r}i}(t), \quad \boldsymbol{x}_{\mathrm{r}i}(0) = \boldsymbol{x}_{i0} \\ \boldsymbol{z}_{\mathrm{r}i}(t) = \boldsymbol{C}_{\mathrm{z}i}\boldsymbol{x}_{\mathrm{r}i}(t) \end{cases} \tag{6.4}$$

an the interconnection relation

$$\boldsymbol{s}_{\mathrm{r}i}(t) = \sum_{j=1}^{N} \boldsymbol{L}_{ij}\boldsymbol{z}_{\mathrm{r}j}(t) \tag{6.5}$$

describes the couplings among the subsystems (6.4).

In line with the previously presented event-based control approaches, the aim of the decentralized event-based controller to be designed is to approximate the behavior of the decentralized reference system (6.3) with adjustable accuracy.

6.2.2 Description of the components

As illustrated in Fig. 6.2 the local control unit F_i consists of the control input generator C_i and the event generator E_i. Both components are described in the following.

Control input generator C_i. The control input generator C_i includes a model of the reference subsystem $\Sigma_{\mathrm{r}i}$ given in (6.4) in order to determine the control input $\boldsymbol{u}_i(t)$ for the time $t \in [t_{k_i}, t_{k_i+1})$:

$$C_i: \begin{cases} \dot{\boldsymbol{x}}_{\mathrm{s}i}(t) = \bar{\boldsymbol{A}}_i\boldsymbol{x}_{\mathrm{s}i}(t) + \boldsymbol{E}_{\mathrm{s}i}\hat{\boldsymbol{s}}_i(t), \quad \boldsymbol{x}_{\mathrm{s}i}(t_{k_i}^+) = \boldsymbol{x}_i(t_{k_i}) \\ \boldsymbol{u}_i(t) = -\boldsymbol{K}_{\mathrm{d}i}\boldsymbol{x}_{\mathrm{s}i}(t). \end{cases} \tag{6.6}$$

$\boldsymbol{x}_{si} \in \mathbb{R}^{n_i}$ denotes the model state which is reset at the event times t_{k_i} $(k_i \in \mathbb{N}_0)$ and $\hat{\boldsymbol{s}}_i \in \mathbb{R}^{q_i}$ is an estimate of the coupling input $\boldsymbol{s}_i(t)$ for the time interval $[t_{k_i}, t_{k_i+1})$. Two different methods for the estimation of the coupling input $\boldsymbol{s}_i(t)$ are presented in Sec. 6.2.5.

Event generator E_i. In order to detect the time instants t_{k_i} at which the model state $\boldsymbol{x}_{\text{s}i}(t)$ in (6.6) needs to be reinitialized, the event generator uses the model

$$E_i : \begin{cases} \dot{\boldsymbol{x}}_{\text{s}i}(t) = \bar{\boldsymbol{A}}_i \boldsymbol{x}_{\text{s}i}(t) + \boldsymbol{E}_{\text{s}i}\hat{\boldsymbol{s}}_i(t), \quad \boldsymbol{x}_{\text{s}i}(t_{k_i}^+) = \boldsymbol{x}_i(t_{k_i}) \\ \quad t_0 = 0, \\ t_{k_i+1} := \inf\left\{t > t_{k_i} \mid |\boldsymbol{x}_i(t) - \boldsymbol{x}_{\text{s}i}(t)| = \bar{\boldsymbol{e}}_i\right\}. \end{cases} \tag{6.7}$$

Here, $\bar{\boldsymbol{e}}_i \in \mathbb{R}_+^{n_i}$ denotes the event threshold vector. The triggering condition is to be understood to hold element-wise, i.e., an event is triggered whenever one element in the vector $|\boldsymbol{x}_i(t) - \boldsymbol{x}_{\text{s}i}(t)|$ equals the corresponding element in $\bar{\boldsymbol{e}}_i$. At the event times t_{k_i} the event generator E_i transmits the current state $\boldsymbol{x}_i(t_{k_i})$ to the control input generator C_i and this information is used in both components to reset the model state $\boldsymbol{x}_{\text{s}i}(t)$. The event generator E_i triggers an initial event at time $t_0 = 0$, regardless of the triggering condition. Thus, by virtue of the event triggering and the reset of the model state $\boldsymbol{x}_{\text{s}i}(t)$ at the event times, the difference state

$$\boldsymbol{x}_{\Delta i}(t) := \boldsymbol{x}_i(t) - \boldsymbol{x}_{\text{s}i}(t)$$

is bounded by

$$\sup_{t\geq 0} |\boldsymbol{x}_{\Delta i}(t)| = \sup_{t\geq 0} |\boldsymbol{x}_i(t) - \boldsymbol{x}_{\text{s}i}(t)| = \bar{\boldsymbol{e}}_i, \quad \forall\, t \geq 0. \tag{6.8}$$

Note that this bound also holds element-wise, i.e., each component of $|\boldsymbol{x}_{\Delta i}(t)|$ is smaller or equal than the corresponding component of the vector $\bar{\boldsymbol{e}}_i$.

6.2.3 Behavior of the decentralized event-based state-feedback loop

Consider the subsystem Σ_i together with the control input generator C_i given in (6.6). For the time $t \in [t_{k_i}, t_{k_i+1})$ the controlled subsystem is represented by the state-space model

$$\begin{pmatrix} \dot{\boldsymbol{x}}_i(t) \\ \dot{\boldsymbol{x}}_{\text{s}i}(t) \end{pmatrix} = \begin{pmatrix} \boldsymbol{A}_i & -\boldsymbol{B}_i\boldsymbol{K}_{\text{d}i} \\ & \bar{\boldsymbol{A}}_i \end{pmatrix} \begin{pmatrix} \boldsymbol{x}_i(t) \\ \boldsymbol{x}_{\text{s}i}(t) \end{pmatrix} + \begin{pmatrix} \boldsymbol{E}_i \\ \end{pmatrix} \boldsymbol{d}_i(t) + \begin{pmatrix} \boldsymbol{E}_{\text{s}i} & \\ & \boldsymbol{E}_{\text{s}i} \end{pmatrix} \begin{pmatrix} \boldsymbol{s}_i(t) \\ \hat{\boldsymbol{s}}_i(t) \end{pmatrix}$$

$$\begin{pmatrix} \boldsymbol{x}_i(t_{k_i}^+) \\ \boldsymbol{x}_{\text{s}i}(t_{k_i}^+) \end{pmatrix} = \begin{pmatrix} \boldsymbol{I}_{n_i} & \boldsymbol{O}_{n_i} \\ \boldsymbol{I}_{n_i} & \boldsymbol{O}_{n_i} \end{pmatrix} \begin{pmatrix} \boldsymbol{x}_i(t_{k_i}) \\ \boldsymbol{x}_{\text{s}i}(t_{k_i}) \end{pmatrix}$$

with the coupling output

$$\boldsymbol{z}_i(t) = \begin{pmatrix} \boldsymbol{C}_{zi} & \boldsymbol{O}_{r_i \times n_i} \end{pmatrix} \begin{pmatrix} \boldsymbol{x}_i(t) \\ \boldsymbol{x}_{si}(t) \end{pmatrix}.$$

In order to investigate the behavior of the difference state $\boldsymbol{x}_{\Delta i}(t)$ the transformation

$$\begin{pmatrix} \boldsymbol{x}_i(t) \\ \boldsymbol{x}_{\Delta i}(t) \end{pmatrix} = \begin{pmatrix} \boldsymbol{I}_{n_i} & \\ \boldsymbol{I}_{n_i} & -\boldsymbol{I}_{n_i} \end{pmatrix} \begin{pmatrix} \boldsymbol{x}_i(t) \\ \boldsymbol{x}_{si}(t) \end{pmatrix}$$

is applied to the former model:

$$\begin{pmatrix} \dot{\boldsymbol{x}}_i(t) \\ \dot{\boldsymbol{x}}_{\Delta i}(t) \end{pmatrix} = \begin{pmatrix} \bar{\boldsymbol{A}}_i & \boldsymbol{B}_i \boldsymbol{K}_{di} \\ & \boldsymbol{A}_i \end{pmatrix} \begin{pmatrix} \boldsymbol{x}_i(t) \\ \boldsymbol{x}_{\Delta i}(t) \end{pmatrix} + \begin{pmatrix} \boldsymbol{E}_i \\ \boldsymbol{E}_i \end{pmatrix} \boldsymbol{d}_i(t) + \begin{pmatrix} \boldsymbol{E}_{si} & \\ \boldsymbol{E}_{si} & -\boldsymbol{E}_{si} \end{pmatrix} \begin{pmatrix} \boldsymbol{s}_i(t) \\ \hat{\boldsymbol{s}}_i(t) \end{pmatrix} \tag{6.9a}$$

$$\begin{pmatrix} \boldsymbol{x}_i(t_{k_i}^+) \\ \boldsymbol{x}_{\Delta i}(t_{k_i}^+) \end{pmatrix} = \begin{pmatrix} \boldsymbol{I}_{n_i} & \boldsymbol{O}_{n_i} \\ \boldsymbol{O}_{n_i} & \boldsymbol{O}_{n_i} \end{pmatrix} \begin{pmatrix} \boldsymbol{x}_i(t_{k_i}) \\ \boldsymbol{x}_{\Delta i}(t_{k_i}) \end{pmatrix} \tag{6.9b}$$

$$\boldsymbol{z}_i(t) = \begin{pmatrix} \boldsymbol{C}_{zi} & \boldsymbol{O}_{r_i \times n_i} \end{pmatrix} \begin{pmatrix} \boldsymbol{x}_i(t) \\ \boldsymbol{x}_{\Delta i}(t) \end{pmatrix}. \tag{6.9c}$$

For the difference state $\boldsymbol{x}_{\Delta i}(t)$ the model (6.9) yields

$$\boldsymbol{x}_{\Delta i}(t) = \int_{t_{k_i}}^{t} \mathrm{e}^{\boldsymbol{A}_i(t-\tau)} \left(\boldsymbol{E}_i \boldsymbol{d}_i(\tau) + \boldsymbol{E}_{si} \left(\boldsymbol{s}_i(\tau) - \hat{\boldsymbol{s}}_i(\tau) \right) \right) \mathrm{d}\tau \tag{6.10}$$

for the time $t \in [t_{k_i}, t_{k_i+1})$. The last result shows that the difference state $\boldsymbol{x}_{\Delta i}(t)$ is directly affected by the disturbance $\boldsymbol{d}_i(t)$, which is due to the fact that no disturbance estimation is used in the generators C_i and E_i. On the other hand, the magnitude of the coupling input $\boldsymbol{s}_i(t)$ is diminished by the estimation $\hat{\boldsymbol{s}}_i(t)$. Considering that an event is triggered by E_i whenever $|\boldsymbol{x}_{\Delta i}(t)| = \bar{\boldsymbol{e}}_i$ holds, it can be seen from (6.10) that the triggering of events can be postponed if the difference between the coupling input $\boldsymbol{s}_i(t)$ and its estimation $\hat{\boldsymbol{s}}_i(t)$ is small.

6.2.4 Approximation of the reference system behavior

In Theorem 4.1 a general condition for the boundedness of the deviation between the behavior of the reference system with continuous state feedback and the event-based state feedback is given. Note that in case of decentralized state feedback, the condition (4.5) can be simplified as

$$\sup_{t\geq 0} \left\| \boldsymbol{K}_i \big(\boldsymbol{x}(t) - \boldsymbol{x}_{\mathrm{s}}^i(t) \big) \right\| = \sup_{t\geq 0} \left\| \boldsymbol{K}_{\mathrm{d}i} \big(\boldsymbol{x}_i(t) - \boldsymbol{x}_{\mathrm{s}i}(t) \big) \right\| < \infty \tag{6.11}$$

for all $i \in \mathcal{N}$, since for the decentralized state-feedback gain (6.2)

$$\boldsymbol{K}_i = \begin{pmatrix} \boldsymbol{O}_{m_i \times n_1} & \dots & \boldsymbol{O}_{m_i \times n_{i-1}} & \boldsymbol{K}_{\mathrm{d}i} & \boldsymbol{O}_{m_i \times n_{i+1}} & \dots & \boldsymbol{O}_{m_i \times n_N} \end{pmatrix}.$$

holds. The condition (6.11) is satisfied by means of the local event triggering and reset of the model state $\boldsymbol{x}_{\mathrm{s}i}(t)$ at the event times. Equations (6.8), (6.11) yield

$$\sup_{t\geq 0} \left\| \boldsymbol{K}_{\mathrm{d}i} \big(\boldsymbol{x}_i(t) - \boldsymbol{x}_{\mathrm{s}i}(t) \big) \right\| \leq \| \boldsymbol{K}_{\mathrm{d}i} \| \cdot \sup_{t\geq 0} \| \boldsymbol{x}_i(t) - \boldsymbol{x}_{\mathrm{s}i}(t) \| = \| \boldsymbol{K}_{\mathrm{d}i} \| \, \| \bar{\boldsymbol{e}}_i \| .$$

Hence, the left-hand side of the relation (6.11) is bounded which implies that the deviation between the behavior of the decentralized reference system (6.3) and the event-based state-feedback loop (2.3), (2.4), (6.6), (6.7) is bounded. The aim of the following analysis is to derive an upper bound for the deviation between both systems.

Consider the interconnected system (2.3), (2.4)

$$\Sigma_i : \begin{cases} \dot{\boldsymbol{x}}_i(t) = \boldsymbol{A}_i \boldsymbol{x}_i(t) + \boldsymbol{B}_i \boldsymbol{u}_i(t) + \boldsymbol{E}_i \boldsymbol{d}_i(t) + \boldsymbol{E}_{\mathrm{s}i} \boldsymbol{s}_i(t), & \boldsymbol{x}_i(0) = \boldsymbol{x}_{0i} \\ \boldsymbol{z}_i(t) = \boldsymbol{C}_{\mathrm{z}i} \boldsymbol{x}_i(t) \end{cases} \tag{6.12a}$$

$$\boldsymbol{s}_i(t) = \sum_{j=1}^{N} \boldsymbol{L}_{ij} \boldsymbol{z}_j(t), \tag{6.12b}$$

subject to the control inputs $\boldsymbol{u}_i(t)$ generated by the control input generators C_i given in (6.6). The plant state $\boldsymbol{x}(t)$ of the controlled overall system is represented by the state-space model

$$\dot{\boldsymbol{x}}(t) = \bar{\boldsymbol{A}} \boldsymbol{x}(t) + \boldsymbol{B} \boldsymbol{K}_{\mathrm{d}} \big(\boldsymbol{x}(t) - \boldsymbol{x}_{\mathrm{s}}(t) \big) + \boldsymbol{E} \boldsymbol{d}(t), \quad \boldsymbol{x}(0) = \boldsymbol{x}_0. \tag{6.13}$$

For the deviation $\boldsymbol{e}(t) = \boldsymbol{x}(t) - \boldsymbol{x}_{\mathrm{r}}(t)$ the model

$$\dot{\boldsymbol{e}}(t) = \bar{\boldsymbol{A}} \boldsymbol{e}(t) + \boldsymbol{B} \boldsymbol{K}_{\mathrm{d}} \big(\boldsymbol{x}(t) - \boldsymbol{x}_{\mathrm{s}}(t) \big), \quad \boldsymbol{e}(0) = \boldsymbol{0}$$

is obtained from Eqs. (6.3), (6.13). With the use of Eq. (6.8) the last equation yields

$$\begin{aligned} |\boldsymbol{e}(t)| &= \left| \int_0^t \mathrm{e}^{\bar{\boldsymbol{A}}(t-\tau)} \boldsymbol{B} \boldsymbol{K}_{\mathrm{d}} \boldsymbol{x}_{\Delta}(\tau) \mathrm{d}\tau \right| \\ &\leq \int_0^\infty \left| \mathrm{e}^{\bar{\boldsymbol{A}}\tau} \boldsymbol{B} \boldsymbol{K}_{\mathrm{d}} \right| \mathrm{d}\tau \cdot \bar{\boldsymbol{e}}, \end{aligned}$$

where $\bar{e} = \begin{pmatrix} \bar{e}_1^\top & \dots & \bar{e}_N^\top \end{pmatrix}^\top$ denotes the event threshold vector for the overall event-based control system.

Theorem 6.1 *The deviation* $e(t) = x(t) - x_r(t)$ *between the state of the interconnected system* (6.12a), (6.12b) *with the event-based state-feedback controllers* (6.6), (6.7) *and the state of the decentralized reference system* (6.3) *is bounded from above by*

$$|e(t)| \leq \int_0^\infty \left| e^{\bar{A}\tau} B K_d \right| d\tau \cdot \bar{e} \tag{6.14}$$

for all $t \geq 0$.

Note that the derived bound (6.14) holds element-wise and, thus, yields a more accurate result than a scalar bound that holds for the norm of the overall state (cf. Theorem 5.1).

Given that the state $x_r(t)$ of the decentralized reference system (6.3) is bounded, the result (6.14) implies the boundedness of the state $x(t)$ of the plant in the event-based state-feedback loop (6.12a), (6.12b), (6.6), (6.7). In other words, the practical stability of the decentralized reference system implies the practical stability of the event-based control system.

6.2.5 Coupling input estimation

This section proposes two different methods for the estimation of the coupling input $s_i(t)$ [12]. The first method is the *static approach* to the coupling input estimation where the signal $s_i(t)$ is assumed to be slightly varying over time and the derived estimates $\hat{s}_i(t) = \hat{s}_i(t_{k_i})$ are constant in between consecutive event times $[t_{k_i}, t_{k_i+1})$. The second method is called the *dynamic approach* to the coupling input estimation. It uses a dynamic model of the behavior of the signal $s_i(t)$ to determine estimates $\hat{s}_i(t)$ which are exponentially decreasing in the time interval $[t_{k_i}, t_{k_i+1})$. For both estimation methods it is assumed that

$$q_i \leq n_i, \quad \forall\, i \in \mathcal{N} \tag{6.15}$$

holds, i.e., the dimension q_i of the coupling input $s_i(t)$ is not larger that the dimension n_i of the state $x_i(t)$. Moreover, for the purpose of deriving the estimation methods the disturbance $d_i(t)$ is assumed to be zero:

$$d_i(t) \equiv 0, \quad \forall\, t \geq 0. \tag{6.16}$$

If, however, the disturbance $d_i(t)$ is non-zero, its disregarding leads to an error in the estimation $\hat{s}_i(t)$ which is discussed subsequently to each estimation method.

Static approach to coupling input estimation. The first estimation approach is based on the assumption that the coupling input is constant in between consecutive events

$$\boldsymbol{s}_i(t) = \bar{\boldsymbol{s}}_i, \quad \text{for } t \in [t_{k_i}, t_{k_i+1}). \tag{6.17}$$

The idea of this estimation method is as follows: At the event time t_{k_i+1} the coupling input estimator determines the constant signal $\bar{\boldsymbol{s}}_i$ which, if it has been affecting subsystem Σ_i in the time interval $t \in [t_{k_i}, t_{k_i+1})$, yields the difference state $\boldsymbol{x}_{\Delta i}(t_{k_i+1})$. This value $\bar{\boldsymbol{s}}_i$ is then used as the estimation $\hat{\boldsymbol{s}}_i(t) = \hat{\boldsymbol{s}}_i(t_{k_i+1})$ for the time interval $t \geq t_{k_i+1}$ until the next event occurs.

Consider that at the event time t_{k_i} the estimate $\hat{\boldsymbol{s}}_i(t) = \hat{\boldsymbol{s}}_i(t_{k_i})$ has been determined and is used as the coupling input estimate in both the control input generator C_i and in the event generator E_i. According to (6.9), the difference state $\boldsymbol{x}_{\Delta i}(t)$ is then described for $t \in [t_{k_i}, t_{k_i+1})$ by the state-space model

$$\dot{\boldsymbol{x}}_{\Delta i}(t) = \boldsymbol{A}_i \boldsymbol{x}_{\Delta i}(t) + \boldsymbol{E}_{\text{s}i}\big(\bar{\boldsymbol{s}}_i - \hat{\boldsymbol{s}}_i(t_{k_i})\big), \quad \boldsymbol{x}_{\Delta i}(t_{k_i}^+) = \boldsymbol{0},$$

where Eqs. (6.16), (6.17) have been taken into account. The last equation yields

$$\begin{aligned}\boldsymbol{x}_{\Delta i}(t) &= \int_{t_{k_i}}^{t} \mathrm{e}^{\boldsymbol{A}_i(t-\tau)} \boldsymbol{E}_{\text{s}i}(\bar{\boldsymbol{s}}_i - \hat{\boldsymbol{s}}_i(t_{k_i}))\mathrm{d}\tau \\ &= \boldsymbol{A}_i^{-1}\left(\mathrm{e}^{\boldsymbol{A}_i(t-t_{k_i})} - \boldsymbol{I}_{n_i}\right)\boldsymbol{E}_{\text{s}i}(\bar{\boldsymbol{s}}_i - \hat{\boldsymbol{s}}_i(t_{k_i})).\end{aligned}$$

At the event time t_{k_i+1} the difference state

$$\boldsymbol{x}_{\Delta i}(t_{k_i+1}) = \boldsymbol{A}_i^{-1}\left(\mathrm{e}^{\boldsymbol{A}_i(t_{k_i+1}-t_{k_i})} - \boldsymbol{I}_{n_i}\right)\boldsymbol{E}_{\text{s}i}(\bar{\boldsymbol{s}}_i - \hat{\boldsymbol{s}}_i(t_{k_i}))$$

is known and used to determine the new estimation according to

$$\hat{\boldsymbol{s}}_i(t_{k_i+1}) := \bar{\boldsymbol{s}}_i = \hat{\boldsymbol{s}}_i(t_{k_i}) + \left(\boldsymbol{A}_i^{-1}\left(\mathrm{e}^{\boldsymbol{A}_i(t_{k_i+1}-t_{k_i})} - \boldsymbol{I}_{n_i}\right)\boldsymbol{E}_{\text{s}i}\right)^{+}\boldsymbol{x}_{\Delta i}(t_{k_i+1}). \tag{6.18}$$

The pseudoinverse in (6.18) exists if assumption (6.15) is satisfied and if the matrix

$$\left(\boldsymbol{A}_i^{-1}\left(\mathrm{e}^{\boldsymbol{A}_i(t_{k_i+1}-t_{k_i})} - \boldsymbol{I}_{n_i}\right)\boldsymbol{E}_{\text{s}i}\right)^{\top}\left(\boldsymbol{A}_i^{-1}\left(\mathrm{e}^{\boldsymbol{A}_i(t_{k_i+1}-t_{k_i})} - \boldsymbol{I}_{n_i}\right)\boldsymbol{E}_{\text{s}i}\right)$$

has full rank.

The following proposition gives an answer to the question how large the estimation error if the assumption (6.16) does not hold and the disturbance $\boldsymbol{d}_i(t)$ is non-zero.

Proposition 6.1 *Consider the coupling input estimation* (6.18) *and assume that the disturbance* $\boldsymbol{d}_i(t)$ *is non-zero. Then the estimation* (6.18) *yields*

$$\hat{\boldsymbol{s}}_i(t_{k_i+1}) := \bar{\boldsymbol{s}}_i(t_{k_i}) + \boldsymbol{\varepsilon}_i(t_{k_i+1})$$

where

$$\boldsymbol{\varepsilon}_i(t_{k_i+1}) = \left(\boldsymbol{A}_i^{-1}\left(\mathrm{e}^{\boldsymbol{A}_i(t_{k_i+1} - t_{k_i})} - \boldsymbol{I}_{n_i}\right)\boldsymbol{E}_{\mathrm{s}i}\right)^+ \int_{t_{k_i}}^{t_{k_i+1}} \mathrm{e}^{\boldsymbol{A}_i(t_{k_i+1} - \tau)}\boldsymbol{E}_i\boldsymbol{d}_i(\tau)\mathrm{d}\tau$$

denotes the estimation error.

Proof. See Appendix B.1. □

This investigation shows the following: The larger the magnitude of the disturbance $\boldsymbol{d}_i(t)$ in the interval $[t_{k_i}, t_{k_i+1})$ is, the larger grows the estimation error $\boldsymbol{\varepsilon}_i(t_{k_i+1})$. Consequently, the application of the estimation method (6.18) should be restricted to the case where the subsystem Σ_i is known to be only marginally disturbed.

The proposed estimation method has low computational cost as new estimations are only determined at the event times. However, this method carries the assumption that the coupling input signal $\boldsymbol{s}_i(t)$ is piecewise constant which might be a rough approximation, particularly if the state of the overall system is far away from the setpoint. To overcome this issue, the next paragraph introduces another estimation method that takes the dynamic behavior of the coupling input signal into account.

Dynamic approach to coupling input estimation. Assume that the behavior of the coupling input $\boldsymbol{s}_i(t)$ is approximately characterized in the interval $[t_{k_i}, t_{k_i+1})$ by the linear state-space model

$$\dot{\boldsymbol{s}}_i(t) = \boldsymbol{A}_{\mathrm{s}i}\boldsymbol{s}_i(t), \quad \boldsymbol{s}_i(t_{k_i}) = \boldsymbol{s}_{i,k_i}. \tag{6.19}$$

Based on the consideration that the closed-loop system is stable the matrix $\boldsymbol{A}_{\mathrm{s}i}$ is assumed to be Hurwitz. The coupling input estimator in C_i and E_i then incorporates the model

$$\frac{\mathrm{d}}{\mathrm{d}t}\hat{\boldsymbol{s}}_i(t) = \boldsymbol{A}_{\mathrm{s}i}\hat{\boldsymbol{s}}_i(t), \quad \hat{\boldsymbol{s}}_i(t_{k_i}^+) = \hat{\boldsymbol{s}}_{i,k_i} \tag{6.20}$$

for $t \in [t_{k_i}, t_{k_i+1})$ in order to determine the estimate $\hat{\boldsymbol{s}}_i(t)$.

The idea of this estimation method is illustrated in Fig. 6.3. Following the same arguments as for the previous estimation method, the aim is to determine at the event time t_{k_i+1} the initial condition $\boldsymbol{s}_i(t_{k_i})$ of the model (6.19) such that the signal $\boldsymbol{s}_i(t)$ for $t \in [t_{k_i}, t_{k_i+1})$ yields the

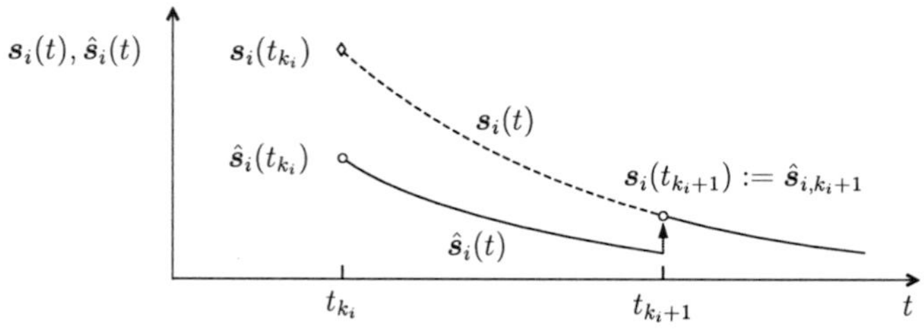

Figure 6.3: Illustration of the dynamic estimation method

difference state $\boldsymbol{x}_{\Delta i}(t_{k_i+1})$ that is detected at the event time t_{k_i+1}. The value $\boldsymbol{s}_i(t_{k_i+1})$ is then applied as initial condition $\hat{\boldsymbol{s}}_i(t_{k_i+1}) = \hat{\boldsymbol{s}}_{i,k_i+1}$ in the model (6.20) to determine the estimate $\hat{\boldsymbol{s}}_i(t)$ for $t \geq t_{k_i+1}$.

To determine $\boldsymbol{s}_i(t_{k_i})$ consider the controlled subsystem Σ_i described by (6.12a), (6.6) together with the coupling input (6.19)

$$\dot{\boldsymbol{x}}_i(t) = \boldsymbol{A}_i \boldsymbol{x}_i(t) - \boldsymbol{B}_i \boldsymbol{K}_{\mathrm{d}i} \boldsymbol{x}_{\mathrm{s}i}(t) + \boldsymbol{E}_{\mathrm{s}i} \boldsymbol{s}_i(t), \qquad \boldsymbol{x}_i(0) = \boldsymbol{x}_{0i} \tag{6.21a}$$

$$\dot{\boldsymbol{s}}_i(t) = \boldsymbol{A}_{\mathrm{s}i} \boldsymbol{s}_i(t), \qquad \boldsymbol{s}_i(0) = \boldsymbol{s}_{i,0} \tag{6.21b}$$

and the control input generator (6.6) with the coupling input (6.20)

$$\dot{\boldsymbol{x}}_{\mathrm{s}i}(t) = \bar{\boldsymbol{A}}_i \boldsymbol{x}_{\mathrm{s}i}(t) + \boldsymbol{E}_{\mathrm{s}i} \hat{\boldsymbol{s}}_i(t), \qquad \boldsymbol{x}_{\mathrm{s}i}(t_{k_i}^+) = \boldsymbol{x}_i(t_k) \tag{6.22a}$$

$$\tfrac{\mathrm{d}}{\mathrm{d}t} \hat{\boldsymbol{s}}_i(t) = \boldsymbol{A}_{\mathrm{s}i} \hat{\boldsymbol{s}}_i(t), \qquad \hat{\boldsymbol{s}}_i(t_{k_i}^+) = \hat{\boldsymbol{s}}_{i,k_i}. \tag{6.22b}$$

which is valid for $t \in [t_{k_i}, t_{k_i+1})$. The difference between the behavior of the systems (6.21) and (6.22) in the interval $[t_{k_i}, t_{k_i+1})$ is represented by the state-space model

$$\frac{\mathrm{d}}{\mathrm{d}t} \begin{pmatrix} \boldsymbol{x}_{\Delta i}(t) \\ \boldsymbol{s}_{\Delta i}(t) \end{pmatrix} = \begin{pmatrix} \boldsymbol{A}_i & \boldsymbol{E}_{\mathrm{s}i} \\ \boldsymbol{O} & \boldsymbol{A}_{\mathrm{s}i} \end{pmatrix} \begin{pmatrix} \boldsymbol{x}_{\Delta i}(t) \\ \boldsymbol{s}_{\Delta i}(t) \end{pmatrix}, \quad \begin{pmatrix} \boldsymbol{x}_{\Delta i}(t_{k_i}^+) \\ \boldsymbol{s}_{\Delta i}(t_{k_i}^+) \end{pmatrix} = \begin{pmatrix} \boldsymbol{0} \\ \boldsymbol{s}_i(t_{k_i}) - \hat{\boldsymbol{s}}_{i,k_i} \end{pmatrix},$$

where $\boldsymbol{s}_{\Delta i}(t) = \boldsymbol{s}_i(t) - \hat{\boldsymbol{s}}_i(t)$. From this model

$$\boldsymbol{x}_{\Delta i}(t) = \begin{pmatrix} \boldsymbol{I}_{n_i} & \boldsymbol{O} \end{pmatrix} \begin{pmatrix} \boldsymbol{x}_{\Delta i}(t) \\ \boldsymbol{s}_{\Delta i}(t) \end{pmatrix} = \begin{pmatrix} \boldsymbol{I}_{n_i} & \boldsymbol{O} \end{pmatrix} \mathrm{e}^{\boldsymbol{R}_i (t - t_{k_i})} \begin{pmatrix} \boldsymbol{x}_{\Delta i}(t_{k_i}^+) \\ \boldsymbol{s}_{\Delta i}(t_{k_i}^+) \end{pmatrix} \tag{6.23}$$

is obtained with

$$\boldsymbol{R}_i = \begin{pmatrix} \boldsymbol{A}_i & \boldsymbol{E}_{si} \\ \boldsymbol{O} & \boldsymbol{A}_{si} \end{pmatrix}. \tag{6.24}$$

Note that the state at time $t_{k_i}^+$ in (6.23) can be expressed as

$$\begin{pmatrix} \boldsymbol{x}_{\Delta i}(t_{k_i}^+) \\ \boldsymbol{s}_{\Delta i}(t_{k_i}^+) \end{pmatrix} = \begin{pmatrix} \boldsymbol{O} \\ \boldsymbol{I}_{q_i} \end{pmatrix} \left(\boldsymbol{s}_i(t_{k_i}) - \hat{\boldsymbol{s}}_{i,k_i}\right)$$

which yields

$$\boldsymbol{x}_{\Delta i}(t) = \begin{pmatrix} \boldsymbol{I}_{n_i} & \boldsymbol{O} \end{pmatrix} \mathrm{e}^{\boldsymbol{R}_i(t - t_{k_i})} \begin{pmatrix} \boldsymbol{O} \\ \boldsymbol{I}_{q_i} \end{pmatrix} \left(\boldsymbol{s}_i(t_{k_i}) - \hat{\boldsymbol{s}}_{i,k}\right).$$

The last equation reflects the relation between the difference $\boldsymbol{s}_i(t_{k_i}) - \hat{\boldsymbol{s}}_{i,k_i}$ and the difference state $\boldsymbol{x}_{\Delta i}(t)$. At the next event time t_{k_i+1} the difference state $\boldsymbol{x}_{\Delta i}(t_{k_i+1})$ is used to determine $\boldsymbol{s}_i(t_{k_i})$:

$$\boldsymbol{s}_i(t_{k_i}) = \hat{\boldsymbol{s}}_{i,k_i} + \left(\bar{\boldsymbol{R}}_i(t_{k_i+1} - t_{k_i})\right)^+ \boldsymbol{x}_{\Delta i}(t_{k_i+1}) \tag{6.25}$$

where

$$\bar{\boldsymbol{R}}_i(t_{k_i+1} - t_{k_i}) = \begin{pmatrix} \boldsymbol{I}_{n_i} & \boldsymbol{O} \end{pmatrix} \mathrm{e}^{\boldsymbol{R}_i(t_{k_i+1} - t_{k_i})} \begin{pmatrix} \boldsymbol{O} \\ \boldsymbol{I}_{q_i} \end{pmatrix}.$$

The pseudoinverse of the matrix $\bar{\boldsymbol{R}}_i(\cdot)$ exists if the matrix $\bar{\boldsymbol{R}}_i^\top \bar{\boldsymbol{R}}_i$ has full rank and relation (6.15) is fulfilled. For the new initial condition of the model (6.20)

$$\begin{aligned} \hat{\boldsymbol{s}}_{i,k_i+1} &:= \mathrm{e}^{\boldsymbol{A}_{si}(t_{k_i+1} - t_{k_i})} \boldsymbol{s}_i(t_{k_i}) \\ &= \mathrm{e}^{\boldsymbol{A}_{si}(t_{k_i+1} - t_{k_i})} \left(\hat{\boldsymbol{s}}_{i,k} + \left(\bar{\boldsymbol{R}}_i(t_{k_i+1} - t_{k_i})\right)^+ \boldsymbol{x}_{\Delta i}(t_{k_i+1})\right) \end{aligned} \tag{6.26}$$

is used. At time $t = 0$ the estimation of the coupling input is initialized with $\hat{\boldsymbol{s}}_{i,0} = \boldsymbol{0}$.

Now consider again that the disturbance $\boldsymbol{d}_i(t)$ is non-zero and, hence, affects subsystem Σ_i. The following proposition shows how large the estimation error in that case is.

Proposition 6.2 *Consider the coupling input estimation* (6.26) *and assume that the disturbance* $\boldsymbol{d}_i(t)$ *is non-zero. Then Eq.* (6.26) *yields*

$$\hat{\boldsymbol{s}}_{i,k_i+1} + \boldsymbol{\varepsilon}_i(t_{k_i+1}) = \mathrm{e}^{\boldsymbol{A}_{\mathrm{s}i}(t_{k_i+1} - t_{k_i})} \boldsymbol{s}_i(t_{k_i}).$$

instead of the initial condition $\hat{\boldsymbol{s}}_i(t_{k_i+1})$, *where*

$$\boldsymbol{\varepsilon}_i(t_{k_i+1}) = \mathrm{e}^{\boldsymbol{A}_{\mathrm{s}i}(t_{k_i+1} - t_{k_i})} \cdot \left(\bar{\boldsymbol{R}}_i(t_{k_i+1} - t_{k_i})\right)^+ \times \int_{t_{k_i}}^{t_{k_i+1}} \begin{pmatrix} \boldsymbol{I}_{n_i} & \boldsymbol{O} \end{pmatrix} \mathrm{e}^{\boldsymbol{R}_i(t_{k_i+1} - \tau)} \begin{pmatrix} \boldsymbol{E}_i \\ \boldsymbol{O} \end{pmatrix} \boldsymbol{d}_i(\tau) \mathrm{d}\tau \quad (6.27)$$

denotes the estimation error.

Proof. See Appendix B.2. □

In summary, the proposed coupling input estimator incorporates the model (6.20). At the event time t_{k_i}, the state of this model is reset according to Eqs. (6.25), (6.26) for which only the difference state $\boldsymbol{x}_{\Delta i}(t_{k_i})$ needs to be known. Note that the dynamics of the coupling input signal $\boldsymbol{s}_i(t)$ is only approximately represented by the autonomous model (6.19). Hence, the matrix $\boldsymbol{A}_{\mathrm{s}i}$ is a design parameter of the proposed estimation method and the choice of this matrix generally depends upon the particular application.

6.2.6 Example: Decentralized event-based control of the thermofluid process

In this example the decentralized event-based state-feedback approach is tested on the thermofluid process, introduced in Sec. 2.4. Both the static as well as the dynamic coupling estimation method are compared in a simulation. Moreover, the decentralized event-based control together with the dynamic coupling estimation method is evaluated in an experiment.

The overall system is composed of two subsystems, which are the reactors TB and TS. The state of both reactors is given by

$$\boldsymbol{x}_1(t) = \begin{pmatrix} l_{\mathrm{B}}(t) \\ \vartheta_{\mathrm{B}}(t) \end{pmatrix}, \quad \boldsymbol{x}_2(t) = \begin{pmatrix} l_{\mathrm{S}}(t) \\ \vartheta_{\mathrm{S}}(t) \end{pmatrix}$$

which refer to the level and the temperature in TB or TS, respectively. For the following investigations the initial state of these subsystems is assumed to be

$$\boldsymbol{x}_1(0) = \begin{pmatrix} 0.05 \\ 5 \end{pmatrix}, \quad \boldsymbol{x}_2(0) = \begin{pmatrix} -0.05 \\ -5 \end{pmatrix} \tag{6.28}$$

which corresponds to initial level in m and temperature in K. In order to concentrate on the effect of the coupling estimation on the control performance, the subsystems are not subject to exogenous disturbances. In the event-based controllers F_i $(i = 1, 2)$ the state-feedback gains

$$\boldsymbol{K}_{\mathrm{d1}} = \begin{pmatrix} 7.28 & 0 \\ 0.89 & -0.08 \end{pmatrix}, \quad \boldsymbol{K}_{\mathrm{d2}} = \begin{pmatrix} 7.73 & 0 \\ 1.12 & 0.05 \end{pmatrix}$$

are applied, which are supposed to yield a desired behavior for continuous state feedback. For the event generators E_i the event thresholds $\bar{\boldsymbol{e}}_1$ and $\bar{\boldsymbol{e}}_2$ are set to

$$\bar{\boldsymbol{e}}_1 = \begin{pmatrix} 0.04 \\ 0.5 \end{pmatrix}, \quad \bar{\boldsymbol{e}}_2 = \begin{pmatrix} 0.04 \\ 0.5 \end{pmatrix}.$$

Decentralized event-based control with static coupling estimation. Figure 6.4 illustrates the results of a simulation that investigates the decentralized event-based state feedback using the static approach to the coupling estimation. From top to bottom the figure shows the level, the temperature, the coupling input signals and the event times for both the reactor TB (left-hand side) and the reactor TS (right-hand side). For both subsystems the coupling input is a two-dimensional signal:

$$\boldsymbol{s}_1(t) = \begin{pmatrix} s_{11}(t) \\ s_{12}(t) \end{pmatrix}, \quad \boldsymbol{s}_2(t) = \begin{pmatrix} s_{21}(t) \\ s_{22}(t) \end{pmatrix}.$$

In the figures in the first two rows of Fig. 6.4 the dashed lines represent the model states

$$\boldsymbol{x}_{\mathrm{s1}}(t) = \begin{pmatrix} l_{\mathrm{sB}}(t) \\ \vartheta_{\mathrm{sB}}(t) \end{pmatrix}, \quad \boldsymbol{x}_2(t) = \begin{pmatrix} l_{\mathrm{sS}}(t) \\ \vartheta_{\mathrm{sS}}(t) \end{pmatrix}$$

which are reset to the corresponding plant state $\boldsymbol{x}_i(t)$ whenever an event is triggered. In the figures in row three and four, the dashed lines show the piecewise constant estimates ($\hat{s}_{11}(t)$, $\hat{s}_{12}(t)$, $\hat{s}_{21}(t)$ and $\hat{s}_{22}(t)$) of the respective coupling input signals ($s_{11}(t)$, $s_{12}(t)$, $s_{21}(t)$ and $s_{22}(t)$).

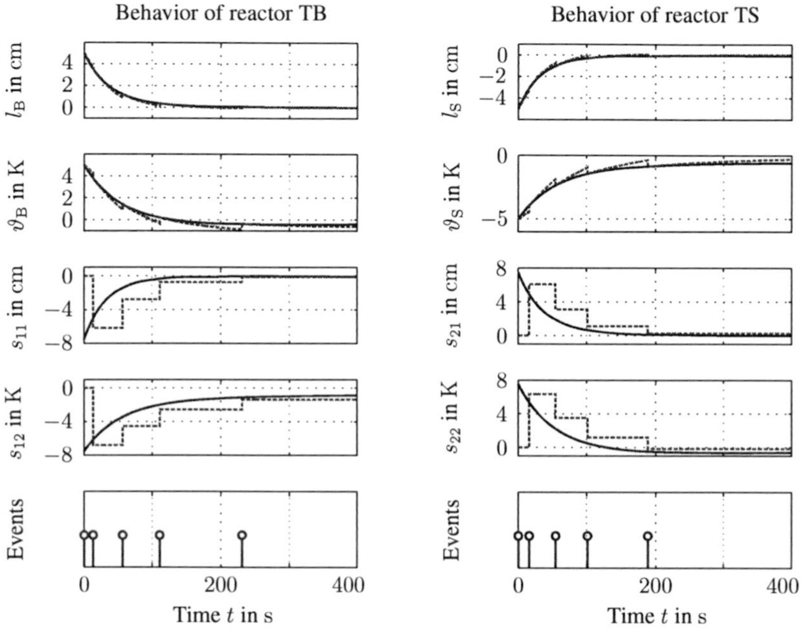

Figure 6.4: Behavior of the decentralized event-based control system with static coupling estimation

In total, 10 events are triggered by the decentralized event-based state-feedback controller until the plant state $\boldsymbol{x}(t)$ is steered from the initial state (6.28) close to the origin. Due to the initial condition (6.28) the magnitudes of the coupling inputs $\boldsymbol{s}_1(t)$ and $\boldsymbol{s}_2(t)$ are large at the beginning of the investigation and get smaller the closer the states converge to the set-point $\bar{\boldsymbol{x}} = \boldsymbol{0}$. Hence, the effect of the interconnection among both subsystems diminishes over time which is also reflected in the triggering of events. In both systems the time that elapses in between consecutive events gets larger until no further event is triggered for $t > 230\,\mathrm{s}$ where the effect of the couplings on both subsystems is marginal. This investigation shows, how the feedback communication is adapted to the system behavior in the decentralized event-based control system.

Figure 6.4 also illustrates the adaption of the estimates $\hat{\boldsymbol{s}}_i$ to the actual coupling signals $\boldsymbol{s}_i(t)$ at the event times t_{k_i}. It can be seen from this figure that the constant estimates only coarsely correspond to the actual trajectories of the coupling inputs $\boldsymbol{s}_i(t)$. The deviation between $\boldsymbol{s}_i(t)$ and the estimation grows with time due to the fact that the constant estimation becomes obsolete

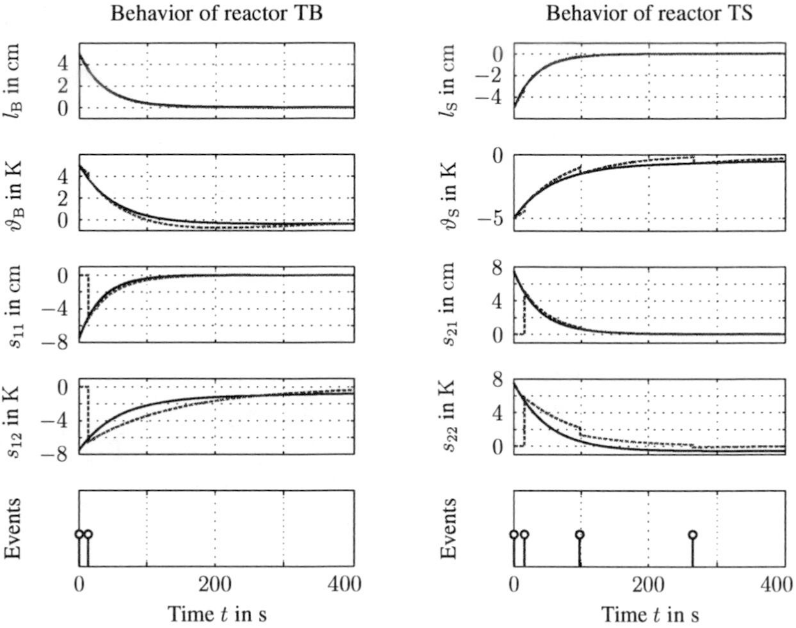

Figure 6.5: Behavior of the decentralized event-based control system with dynamic coupling estimation

which eventually leads to the triggering of a new event. The next part shows how the triggering of events can be further reduced by means of the dynamic coupling estimation.

Decentralized event-based control with dynamic coupling estimation. In the following the results of the simulation of the event-based control system using the dynamic coupling estimation is presented. In order to perform the dynamic approach to the coupling signal estimation, in each control input generator C_i and each event generator E_i the model (6.20) with the respective matrices

$$\boldsymbol{A}_{s1} = \boldsymbol{A}_2 - \boldsymbol{B}_2\boldsymbol{K}_{d2} = 10^{-3}\begin{pmatrix} -25 & 0 \\ 0 & -7.3 \end{pmatrix},$$

$$\boldsymbol{A}_{s2} = \boldsymbol{A}_1 - \boldsymbol{B}_1\boldsymbol{K}_{d1} = 10^{-3}\begin{pmatrix} -22.5 & 0 \\ 0.1 & -11.7 \end{pmatrix}$$

is implemented.

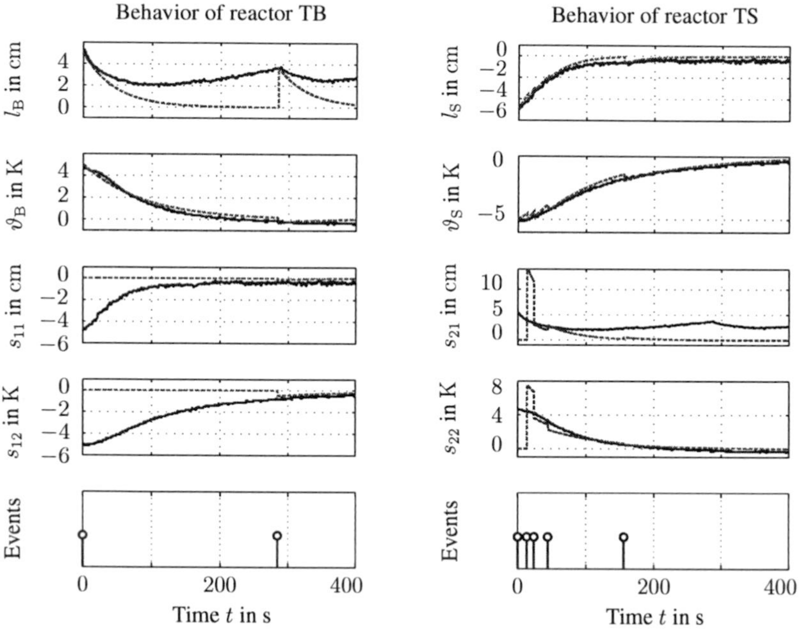

Figure 6.6: Experimental result: Behavior of the event-based control system with dynamic coupling estimation

Figure 6.5 depicts the behavior of the decentralized event-based control system where the event-based controllers use the dynamic coupling estimation. The signals that are shown in this figure are the same as in Fig. 6.4. In comparison with the behavior of the decentralized event-based control system with the static coupling estimation it can be seen in Fig. 6.5 that the triggering of events is considerably reduced. Within the investigated time interval $[0, 400]$ s only 6 events are triggered. Interestingly, in reactor TB only one event (despite the initial event) is triggered. After this event the estimate $\hat{s}_1(t)$ remains close enough to the actual coupling input $s_1(t)$ such that no further feedback communication is required. This investigation shows that the feedback communication effort can be reduced by using a more sophisticated estimation method (compared to the static approach to the coupling estimation) which, however, requires a higher computational effort.

Figure 6.6 shows the result of an experiment in which the decentralized event-based state feedback with the dynamic approach to the coupling estimation is tested on the thermofluid process. The behavior of the real system only slightly deviates from the results shown in Fig. 6.5

that are obtained by means of a simulation. The differences between both results can be ascribed to model uncertainties. First, the event that is triggered at $t = 390\,\mathrm{s}$ by E_1 is caused by the deviation of the level $l_{\mathrm{B}}(t)$ from the model state $l_{\mathrm{sB}}(t)$. This deviation occurs, because at the technical plant the inflow into and the outflow out of reactor TB are not equal. Hence, the level $l_{\mathrm{B}}(t)$ increases and leads to the triggering of an event which, thus, is not due to the interconnection to reactor TS. In reactor TS the estimate $\hat{\boldsymbol{s}}_2(t)$ that is determined at the first event after the initial one substantially deviates from the actual coupling input $\boldsymbol{s}_2(t)$ which can also be explained by uncertainties of the model used in the control input generator C_2 and in the event generator E_2. Nevertheless, the experimental results show that the decentralized event-based state feedback works well together with the dynamic coupling input signal estimation and, moreover, is robust with respect to model uncertainties.

6.3 Analysis of interconnected event-based state-feedback loops

The main part of this section is concerned with the stability analysis of N interconnected event-based state-feedback loops (Fig. 6.7). The event-based controller F_i for each subsystem Σ_i is designed according to the approach presented in Ch. 3, assuming that Σ_i is not interconnected to any other subsystem. Hence, the N isolated event-based state-feedback loops (with $\boldsymbol{L} = \boldsymbol{O}$) are stable (cf. Theorem 3.1). The central question that is to be answered by the analysis in this section is:

> How strong can the interconnection (6.12b) among the subsystems (6.12a) be, such that the stability of the isolated event-based state-feedback loops implies the stability of the overall event-based control system?

In this regard, the strength of the interconnection refers to the magnitude of the elements in $\boldsymbol{L}$ and in this section it is analyzed how large $\boldsymbol{L}$ can grow such that the stability of the overall control system is still retained [11, 12].

The investigation presented in this section differs from the one in Sec. 6.2 in that a decentralized design of the local control units F_i is explicitly considered here and the main analysis problem is to find a condition that guarantees the stability of the interconnected control loops. In contrast to that, the previous section assumes a stabilizing decentralized state feedback gain $\boldsymbol{K}_\mathrm{d}$ for the overall system to be given and presents a method for its implementation in an event-based fashion.

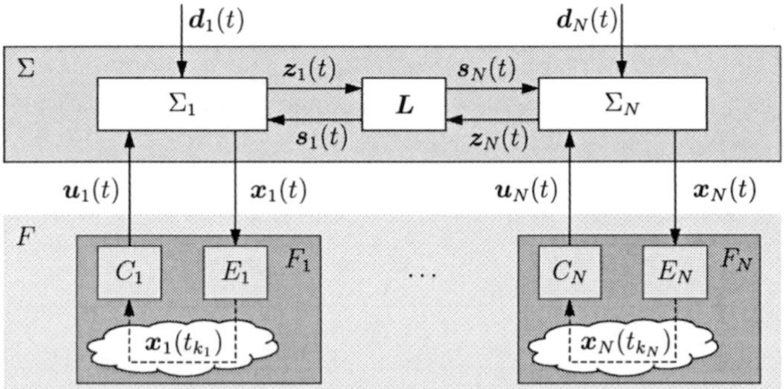

Figure 6.7: Interconnection of N event-based state-feedback loops

The main results of the subsequent analysis are the following:

- A condition on the interconnection matrix $\boldsymbol{L}$ is derived that can be used to test the stability of the overall event-based control system, given that the isolated event-based state-feedback loops are known to be stable (Theorem. 6.2).

- The event-based control system is shown to be practically stable with respect to the set $\mathcal{A}$ (Theorem 6.4). The size of the set $\mathcal{A}$ is expressed in form of a vectorial bound that holds element-wise for each component of the state $\boldsymbol{x}(t)$ and, thus, is less conservative than a bound on the norm of the state which is usually proposed in literature, e.g., in [54, 98, 115].

- A lower bound on the minimum inter-event time for events generated by one event generator E_i is presented (Theorem 6.6).

6.3.1 Structure of the event-based control system

The event-based state-feedback approach [115] presented in Ch. 3 is applied to design the decentralized event-based state-feedback controllers F_i for the isolated subsystems

$$\dot{\boldsymbol{x}}_i(t) = \boldsymbol{A}_i \boldsymbol{x}_i(t) + \boldsymbol{B}_i \boldsymbol{u}_i(t) + \boldsymbol{E}_i \boldsymbol{d}_i(t), \quad \boldsymbol{x}_i(0) = \boldsymbol{x}_{i0}, \tag{6.29}$$

where $\boldsymbol{s}_i(t) \equiv 0$ is assumed for all $t \geq 0$ and all $i \in \mathcal{N}$. First, a continuous state-feedback controller

$$\boldsymbol{u}_i(t) = -\boldsymbol{K}_{\mathrm{d}i} \boldsymbol{x}_i(t) \tag{6.30}$$

is determined which ensures that the resulting reference system

$$\Sigma_{\mathrm{r}i}: \quad \dot{\boldsymbol{x}}_{\mathrm{r}i}(t) = \underbrace{(\boldsymbol{A}_i - \boldsymbol{B}_i \boldsymbol{K}_{\mathrm{d}i})}_{:= \bar{\boldsymbol{A}}_i} \boldsymbol{x}_{\mathrm{r}i}(t) + \boldsymbol{E}_i \boldsymbol{d}_i(t), \quad \boldsymbol{x}_{\mathrm{r}i}(0) = \boldsymbol{x}_{0i} \tag{6.31}$$

has a desired disturbance behavior. As before, the index r is used to distinguish the signals of the reference system (6.31) from the ones that occur in the event-based control systems. Next, the control input generator C_i and event generator E_i of the controller F_i should be designed such that the resulting event-based control loop approximates the behavior of the reference system (6.31) with adjustable precision. According to [115], these components operate as follows.

Control input generator C_i. For the time $t \in [t_{k_i}, t_{k_i+1})$ the control input generator C_i uses a model of the reference system (6.31) to generate the control input $\boldsymbol{u}_i(t)$:

$$C_i : \begin{cases} \Sigma_{\text{si}} : & \dot{\boldsymbol{x}}_{\text{si}}(t) = \bar{\boldsymbol{A}}_i \boldsymbol{x}_{\text{si}}(t) + \boldsymbol{E}_i \hat{\boldsymbol{d}}_{i,k_i}, \quad \boldsymbol{x}_{\text{si}}(t_{k_i}^+) = \boldsymbol{x}_i(t_{k_i}) \\ & \boldsymbol{u}_i(t) = -\boldsymbol{K}_i \boldsymbol{x}_{\text{si}}(t). \end{cases} \tag{6.32}$$

$\boldsymbol{x}_{\text{si}} \in \mathbb{R}^{n_i}$ and $\hat{\boldsymbol{d}}_{i,k_i} \in \mathbb{R}^{p_i}$ denote the model state and an estimation of the disturbance $\boldsymbol{d}_i(t)$, respectively. The estimation method given in Sec. 3.2.3 can be easily adapted to determine the estimates $\hat{\boldsymbol{d}}_{i,k_i}$ of the local disturbance $\boldsymbol{d}_i(t)$. Following the same idea of this estimation method yields

$$\hat{\boldsymbol{d}}_{i,0} = \boldsymbol{0} \tag{6.33a}$$

$$\hat{\boldsymbol{d}}_{i,k_i} = \hat{\boldsymbol{d}}_{i,k_i-1} + \left(\boldsymbol{A}_i^{-1}\left(\mathrm{e}^{\boldsymbol{A}_i(t_{k_i} - t_{k_i-1})} - \boldsymbol{I}_{n_i}\right)\boldsymbol{E}_i\right)^+ \left(\boldsymbol{x}_i(t_{k_i}) - \boldsymbol{x}_{\text{si}}(t_{k_i})\right) \tag{6.33b}$$

where $p_i \leq n_i$ is presumed, i.e., the dimension p_i of the disturbance $\boldsymbol{d}_i(t)$ does not exceed the dimension n_i of the subsystem state $\boldsymbol{x}_i(t)$.

Event generator E_i. The event generator E_i is described by the model

$$E_i : \begin{cases} \Sigma_{\text{si}} : & \dot{\boldsymbol{x}}_{\text{si}}(t) = \bar{\boldsymbol{A}}_i \boldsymbol{x}_{\text{si}}(t) + \boldsymbol{E}_i \hat{\boldsymbol{d}}_{i,k_i}, \quad \boldsymbol{x}_{\text{si}}(t_{k_i}^+) = \boldsymbol{x}_i(t_{k_i}) \\ & t_0 = 0, \\ & t_{k_i+1} := \inf\left\{t > t_{k_i} \mid |\boldsymbol{x}_i(t) - \boldsymbol{x}_{\text{si}}(t)| = \bar{\boldsymbol{e}}_i\right\}. \end{cases} \tag{6.34}$$

In contrast to the centralized event-based state-feedback approach [115], a vectorial event threshold $\bar{\boldsymbol{e}}_i \in \mathbb{R}_+^{n_i}$ is applied here in the triggering condition instead of a scalar event threshold. Hence, E_i triggers an event whenever the absolute value of one component of the difference state

$$\boldsymbol{x}_{\Delta i}(t) = \boldsymbol{x}_i(t) - \boldsymbol{x}_{\text{si}}(t)$$

equals the corresponding element in $\bar{\boldsymbol{e}}_i$. At the event time t_{k_i}, E_i transmits the current subsystem state $\boldsymbol{x}_i(t_{k_i})$ to C_i and both generators use this information to reset the state $\boldsymbol{x}_{\text{si}}(t)$ of the model Σ_{si}. Note that E_i triggers an initial event at $t = 0$. Hence, the difference state $\boldsymbol{x}_{\Delta i}(t)$ is bounded by

$$|\boldsymbol{x}_{\Delta i}(t)| = |\boldsymbol{x}_i(t) - \boldsymbol{x}_{\text{si}}(t)| \leq \bar{\boldsymbol{e}}_i, \quad \forall\, t \geq 0 \tag{6.35}$$

due to the event triggering and the state reset at the event times. For the disturbance estimation the event generator E_i also applies the method (6.33).

Note that both the control input generator C_i and the event generator E_i neglect the influence of the interconnection (6.12b) of Σ_i to other subsystems. Hence, the overall event-based control system (6.12a), (6.12b), (6.32), (6.34) is stable if $\boldsymbol{L} = \boldsymbol{O}$ holds. The following analysis derives a condition on the matrix $\boldsymbol{L}$ which guarantees that the event-based control loops with non-vanishing interconnection retain stable.

6.3.2 Basic idea of the stability analysis

This section explains the underlying idea of the stability analysis of the interconnected event-based state-feedback loops (6.12a), (6.12b), (6.32), (6.34) that is presented in the following. Note that the overall event-based control system is a hybrid dynamical system which exhibits a complex behavior, characterized by a sequence of state jumps at the event times (at time t_{k_i} the model state $\boldsymbol{x}_{\mathrm{s}i}(t_{k_i}^+)$ "jumps" to the value of the subsystem state $\boldsymbol{x}_i(t_{k_i})$) and continuous dynamics in between events [5, 54]. Moreover, the state jumps generally occur asynchronously in time. From this consideration follows that the investigation of the overall system behavior in between consecutive events, as done in the work [115], would lead to a very complicated analysis method.

In order to reduce this complexity the subsequently presented analysis first develops comparison systems for the event-based control loops. These comparison systems are linear systems that produce upper bounds on the signals that occur in the event-based control loops for all $t \geq 0$. Second, a condition for the stability of the interconnected comparison systems is derived which guarantees that the signals of these comparison systems remain bounded for all time $t \geq 0$ which implies the boundedness of the signals of the interconnected event-based control loops. Hence, the stability condition to be derived for the interconnection of the comparison systems can also be used to test stability of the interconnection of the event-based control loops.

6.3.3 Comparison systems

The stability analysis in this section makes use of comparison systems which yield upper bounds on the signals of the respective subsystems.

Consider the isolated subsystem (6.29) that has the state trajectory

$$\boldsymbol{x}_i(t) = \mathrm{e}^{\bar{\boldsymbol{A}}_i t}\boldsymbol{x}_{0i} + \boldsymbol{G}_{\mathrm{xu}i} * \boldsymbol{u}_i + \boldsymbol{G}_{\mathrm{xd}i} * \boldsymbol{d}_i, \quad \forall\, t \geq 0$$

where

$$G_{xui}(t) = e^{\bar{A}_i t} B_i, \tag{6.36a}$$

$$G_{xdi}(t) = e^{\bar{A}_i t} E_i \tag{6.36b}$$

denote the impulse response matrices describing the impact of the control input $u_i(t)$ or the disturbance $d_i(t)$, respectively, on the state $x_i(t)$.

Definition 6.1 *The system*

$$r_{xi}(t) = \bar{F}_i(t) |x_{0i}| + \bar{G}_{xui} * |u_i| + \bar{G}_{xdi} * |d_i| \tag{6.37}$$

with $r_{xi} \in \mathbb{R}^{n_i}$ *is called a* comparison system *of subsystem* (6.29) *if it satisfies the inequality*

$$r_{xi}(t) \geq |x_i(t)|, \quad \forall\, t \geq 0$$

for arbitrary but bounded inputs $u_i(t)$ *and* $d_i(t)$.

A method for finding a comparison system is given in the following lemma.

Lemma 6.1 (Lemma 8.3 in [113]) *The system* (6.37) *is a* comparison system *of the system* (6.29) *if and only if the matrix* $\bar{F}_i(t)$ *and the impulse response matrices* $\bar{G}_{xui}(t)$ *and* $\bar{G}_{xdi}(t)$ *satisfy the relations*

$$\bar{F}_i(t) \geq \left| e^{\bar{A}_i t} \right|, \quad \bar{G}_{xui}(t) \geq |G_{xui}(t)|, \quad \bar{G}_{xdi}(t) \geq |G_{xdi}(t)|, \quad \forall\, t \geq 0$$

where the $G_{xui}(t)$ *and* $G_{xdi}(t)$ *are defined in Eq.* (6.36).

6.3.4 Stability of the interconnected event-based state-feedback loops

This section proposes a new stability analysis method for the interconnected event-based state-feedback loops (6.12a), (6.12b), (6.32), (6.34). This analysis method is derived using the comparison principle which leads to a condition that depends upon the interconnection matrix L. The main analysis result is presented in the following theorem.

Theorem 6.2 *Consider the interconnected event-based state-feedback loop (6.12a), (6.12b), (6.32), (6.34) and define*

$$\bar{G}_{xs}(t) := \operatorname{diag}\left(\left|e^{\bar{A}_1 t} E_{s1}\right|, \ldots, \left|e^{\bar{A}_N t} E_{sN}\right|\right), \tag{6.38}$$

$$C_z := \operatorname{diag}\left(C_{z1}, \ldots, C_{zN}\right). \tag{6.39}$$

The plant state $x(t)$ of the decentralized event-based state-feedback loop is bounded for all $t \geq 0$ if the relation

$$\lambda_p\left(\int_0^\infty \bar{G}_{xs}(t)\mathrm{d}t\,|L|\,|C_z|\right) < 1 \tag{6.40}$$

is satisfied.

The proof of Theorem 6.2 is given subsequently to the following discussion of this analysis result.

Consider the relation (6.40) and note that it can be used to test the stability of the interconnected event-based state-feedback loops (6.12a), (6.12b), (6.32), (6.34) for a given interconnection matrix L. Moreover, this condition can also be adapted in order to identify a bound on the coupling strength for which the stability of the overall control system is implied by the stability of the isolated event-based state-feedback loops. To see this, assume that the coupling strength can be adjusted by means of a scaling factor $\kappa \in \mathbb{R}_{\geq 0}$, i.e., the coupling input $s_i(t)$ to subsystem Σ_i $(i \in \mathcal{N})$ is given by

$$s_i(t) = \kappa \sum_{j=1}^{N} L_{ij} z_j(t).$$

rather than by Eq. (6.12b). Then the relation (6.40) can be restated as

$$\kappa < \left(\lambda_p\left(\int_0^\infty \bar{G}_{xs}(t)\mathrm{d}t\,|L|\,|C_z|\right)\right)^{-1} \tag{6.41}$$

which defines an upper bound on the scaling factor κ for which stability of the overall control system can still be guaranteed. This consideration implies that Theorem 6.2 can be regarded as a small-gain condition which is satisfied only for a sufficiently weak interconnection (6.12b) between the event-based state-feedback loops (6.12a), (6.32), (6.34).

Proof. For the stability analysis of the overall event-based control system, first consider the subsystem Σ_i, represented by (6.12a), together with the event-based controller (6.32), (6.34). The controlled subsystem is described by the state-space model

$$\bar{\Sigma}_i : \begin{cases} \dot{\boldsymbol{x}}_i(t) = \bar{\boldsymbol{A}}_i \boldsymbol{x}_i(t) + \boldsymbol{B}_i \boldsymbol{K}_{\mathrm{d}i} \boldsymbol{x}_{\Delta i}(t) + \boldsymbol{E}_i \boldsymbol{d}_i(t) + \boldsymbol{E}_{\mathrm{s}i} \boldsymbol{s}_i(t), \quad \boldsymbol{x}_i(0) = \boldsymbol{x}_{0i} \\ \boldsymbol{z}_i(t) = \boldsymbol{C}_{\mathrm{z}i} \boldsymbol{x}_i(t). \end{cases}$$

The previous model yields the state trajectory

$$\boldsymbol{x}_i(t) = \boldsymbol{F}_i(t) \boldsymbol{x}_{0i} + \boldsymbol{G}_{\mathrm{xx}i} * \boldsymbol{x}_{\Delta i} + \boldsymbol{G}_{\mathrm{xd}i} * \boldsymbol{d}_i + \boldsymbol{G}_{\mathrm{xs}i} * \boldsymbol{s}_i$$

with the matrices

$$\boldsymbol{F}_i(t) = \mathrm{e}^{\bar{\boldsymbol{A}}_i t}, \tag{6.42a}$$

$$\boldsymbol{G}_{\mathrm{xx}i}(t) = \mathrm{e}^{\bar{\boldsymbol{A}}_i t} \boldsymbol{B}_i \boldsymbol{K}_i, \tag{6.42b}$$

$$\boldsymbol{G}_{\mathrm{xd}i}(t) = \mathrm{e}^{\bar{\boldsymbol{A}}_i t} \boldsymbol{E}_i \tag{6.42c}$$

and

$$\boldsymbol{G}_{\mathrm{xs}i}(t) = \mathrm{e}^{\bar{\boldsymbol{A}}_i t} \boldsymbol{E}_{\mathrm{s}i}. \tag{6.42d}$$

An upper bound on the state $\boldsymbol{x}_i(t)$ is obtained by means of the comparison system

$$\boldsymbol{r}_{\mathrm{x}i}(t) = \bar{\boldsymbol{F}}_i(t) |\boldsymbol{x}_{i0}| + \bar{\boldsymbol{G}}_{\mathrm{xx}i} * |\boldsymbol{x}_{\Delta i}| + \bar{\boldsymbol{G}}_{\mathrm{xd}i} * |\boldsymbol{d}_i| + \bar{\boldsymbol{G}}_{\mathrm{xs}i} * |\boldsymbol{s}_i| \geq |\boldsymbol{x}_i(t)|, \quad \forall\, t \geq 0 \tag{6.43}$$

with

$$\bar{\boldsymbol{F}}_i(t) = |\boldsymbol{F}_i(t)|, \quad \bar{\boldsymbol{G}}_{\mathrm{xx}i}(t) = |\boldsymbol{G}_{\mathrm{xx}i}(t)|, \quad \bar{\boldsymbol{G}}_{\mathrm{xd}i} = |\boldsymbol{G}_{\mathrm{xd}i}(t)|, \quad \bar{\boldsymbol{G}}_{\mathrm{xs}i}(t) = |\boldsymbol{G}_{\mathrm{xs}i}(t)|. \tag{6.44}$$

Recall that the difference state $\boldsymbol{x}_{\Delta i}(t)$ is bounded according to Eq. (6.35). Hence, the relation

$$\bar{\boldsymbol{G}}_{\mathrm{xx}i} * |\boldsymbol{x}_{\Delta i}| = \int_0^t \bar{\boldsymbol{G}}_{\mathrm{xx}i}(t-\tau) |\boldsymbol{x}_{\Delta i}(\tau)| \,\mathrm{d}\tau \leq \int_0^\infty \bar{\boldsymbol{G}}_{\mathrm{xx}i}(t) \mathrm{d}t \cdot \bar{\boldsymbol{e}}_i =: \boldsymbol{e}_{\Delta i} \tag{6.45}$$

holds. With the bound (6.45) the comparison system (6.43) can be reformulated as

$$\boldsymbol{r}_{\mathrm{x}i}(t) = \bar{\boldsymbol{F}}_i(t) |\boldsymbol{x}_{i0}| + \boldsymbol{e}_{\Delta i} + \bar{\boldsymbol{G}}_{\mathrm{xd}i} * |\boldsymbol{d}_i| + \bar{\boldsymbol{G}}_{\mathrm{xs}i} * |\boldsymbol{s}_i| \geq |\boldsymbol{x}_i(t)|, \quad \forall\, t \geq 0. \tag{6.46}$$

The following investigation considers the behavior of the overall event-based control system for which a comparison system is derived from (6.46) as follows. First, define

$$\boldsymbol{r}_{\mathrm{x}}(t) := \begin{pmatrix}\boldsymbol{r}_{\mathrm{x}1}^{\top}(t) & \dots & \boldsymbol{r}_{\mathrm{x}N}^{\top}(t)\end{pmatrix}^{\top},$$
$$\boldsymbol{e}_{\Delta} := \begin{pmatrix}\boldsymbol{e}_{\Delta 1}^{\top} & \dots & \boldsymbol{e}_{\Delta N}^{\top}\end{pmatrix}^{\top}$$

and

$$\begin{aligned}\bar{\boldsymbol{F}}(t) &:= \operatorname{diag}\left(\bar{\boldsymbol{F}}_1(t), \dots, \bar{\boldsymbol{F}}_N(t)\right),\\ \bar{\boldsymbol{G}}_{\mathrm{xd}}(t) &:= \operatorname{diag}\left(\bar{\boldsymbol{G}}_{\mathrm{xd}1}(t), \dots, \bar{\boldsymbol{G}}_{\mathrm{xd}N}(t)\right),\\ \bar{\boldsymbol{G}}_{\mathrm{xs}}(t) &:= \operatorname{diag}\left(\bar{\boldsymbol{G}}_{\mathrm{xs}1}(t), \dots, \bar{\boldsymbol{G}}_{\mathrm{xs}N}(t)\right),\\ \boldsymbol{C}_{\mathrm{z}} &:= \operatorname{diag}\left(\boldsymbol{C}_{\mathrm{z}1}, \dots, \boldsymbol{C}_{\mathrm{z}N}\right).\end{aligned} \tag{6.47}$$

The interconnection of the comparison systems (6.46) according to

$$|\boldsymbol{s}(t)| \le |\boldsymbol{L}|\,|\boldsymbol{z}(t)| \le |\boldsymbol{L}|\,|\boldsymbol{C}_{\mathrm{z}}|\,|\boldsymbol{x}(t)|$$

yields the comparison system

$$\begin{aligned}\boldsymbol{r}_{\mathrm{x}}(t) &= \bar{\boldsymbol{F}}(t)\,|\boldsymbol{x}_0| + \boldsymbol{e}_{\Delta} + \bar{\boldsymbol{G}}_{\mathrm{xd}} * |\boldsymbol{d}| + \bar{\boldsymbol{G}}_{\mathrm{xs}} * |\boldsymbol{s}|\\ &= \bar{\boldsymbol{F}}(t)\,|\boldsymbol{x}_0| + \boldsymbol{e}_{\Delta} + \bar{\boldsymbol{G}}_{\mathrm{xd}} * |\boldsymbol{d}| + \bar{\boldsymbol{G}}_{\mathrm{xs}}\,|\boldsymbol{L}|\,|\boldsymbol{C}_{\mathrm{z}}| * |\boldsymbol{x}| \ge |\boldsymbol{x}(t)|, \quad \forall\, t \ge 0\end{aligned} \tag{6.48}$$

for the overall event-based state-feedback system (6.12a), (6.12b), (6.32), (6.34). Note that the relation (6.48) is an implicit bound on the overall plant state $\boldsymbol{x}(t)$. The comparison principle says that if the condition (6.40) is satisfied, the bound (6.48) can be restated as

$$\boldsymbol{r}_{\mathrm{x}}(t) = \boldsymbol{G} * \left(\bar{\boldsymbol{F}}(t)\,|\boldsymbol{x}_0| + \boldsymbol{e}_{\Delta} + \bar{\boldsymbol{G}}_{\mathrm{xd}} * |\boldsymbol{d}|\right) \ge |\boldsymbol{x}(t)|, \quad \forall\, t \ge 0 \tag{6.49}$$

with the new impulse response matrix

$$\boldsymbol{G}(t) = \delta(t)\boldsymbol{I}_n + \bar{\boldsymbol{G}}_{\mathrm{xs}}\,|\boldsymbol{L}|\,|\boldsymbol{C}_{\mathrm{z}}| * \boldsymbol{G}, \tag{6.50}$$

for which the relation

$$\int_0^{\infty} \boldsymbol{G}(t)\mathrm{d}t < \infty \tag{6.51}$$

holds. To see this, consider Eq. (6.50) which yields

$$\int_0^\infty \boldsymbol{G}(t)\mathrm{d}t = \boldsymbol{I}_n + \int_0^\infty \bar{\boldsymbol{G}}_{\mathrm{xs}}(t)\mathrm{d}t\,|\boldsymbol{L}|\,|\boldsymbol{C}_{\mathrm{z}}| \int_0^\infty \boldsymbol{G}(t)\mathrm{d}t.$$

The last equation can be reformulated as

$$\left(\boldsymbol{I}_n - \int_0^\infty \bar{\boldsymbol{G}}_{\mathrm{xs}}(t)\mathrm{d}t\,|\boldsymbol{L}|\,|\boldsymbol{C}_{\mathrm{z}}|\right) \int_0^\infty \boldsymbol{G}(t)\mathrm{d}t = \boldsymbol{I}_n$$

from which the matrix

$$\int_0^\infty \boldsymbol{G}(t)\mathrm{d}t = \left(\boldsymbol{I}_n - \int_0^\infty \bar{\boldsymbol{G}}_{\mathrm{xs}}(t)\mathrm{d}t\,|\boldsymbol{L}|\,|\boldsymbol{C}_{\mathrm{z}}|\right)^{-1} := \hat{\boldsymbol{G}}^{-1} \tag{6.52}$$

follows. In order to proof (6.51) to be true, the matrix $\hat{\boldsymbol{G}}^{-1}$ is now shown to be non-negative. To this end consider Theorem 2.3, which states that

$$\hat{\boldsymbol{G}} = \boldsymbol{I}_n - \int_0^\infty \bar{\boldsymbol{G}}_{\mathrm{xs}}(t)\mathrm{d}t\,|\boldsymbol{L}|\,|\boldsymbol{C}_{\mathrm{z}}|$$

is an M-matrix if and only if condition (6.40) is satisfied. Then, by virtue of Theorem 2.2, the matrix $\hat{\boldsymbol{G}}^{-1}$ is non-negative which implies that (6.51) holds true.

Now consider again the comparison system (6.49) and note that the term $\bar{\boldsymbol{F}}(t)\,|\boldsymbol{x}_0|$ vanishes with time whereas the constant $\boldsymbol{e}_\Delta$ and the disturbance $\boldsymbol{d}(t)$ are non-vanishing but bounded. Given that Eq. (6.51) holds, the signal $\boldsymbol{r}_{\mathrm{x}}(t)$ remains bounded for all $t \geq 0$ and, hence, the plant state $\boldsymbol{x}(t)$ of the event-based control system (6.12a), (6.12b), (6.32), (6.34) is bounded, as well, which completes the proof. □

The condition presented in Theorem 6.2 leads to the following stability test procedure.

Algorithm 6.1 (Stability test for the interconnected event-based state-feedback loops)

Given: Interconnected system (6.12a), (6.12b) with event-based state-feedback controllers (6.32), (6.34) that are designed such that $\bar{\boldsymbol{A}}_i = (\boldsymbol{A}_i - \boldsymbol{B}_i\boldsymbol{K}_{\mathrm{d}i})$ is Hurwitz for all $i \in \mathcal{N}$.

1. Determine the matrices $|\boldsymbol{C}_{\mathrm{z}}|$ and $|\boldsymbol{L}|$.
2. Determine the matrix $\int_0^\infty \bar{\boldsymbol{G}}_{\mathrm{xs}}(t)\mathrm{d}t$ with (6.42), (6.44), (6.47).
3. Check whether the condition (6.40) is satisfied.

Result: If condition (6.40) holds, then the practical stability of the isolated event-based state-feedback loops implies the practical stability of the overall event-based control system (6.12a), (6.12b), (6.32), (6.34).

Note that the stability condition (6.40) is sufficient but not necessary for the stability of the interconnected event-based state-feedback loops (6.12a), (6.12b), (6.32), (6.34). That means, the overall event-based control system might be stable even if the condition (6.40) is not satisfied. The conservatism of this analysis method is introduced by using absolute values of the matrices in (6.44) when building the comparison systems (6.43). In particular, the analysis method yields non-conservative results if

$$\boldsymbol{G}_{\mathrm{xs}i}(t) = |\boldsymbol{G}_{\mathrm{xs}i}(t)|\,, \quad \boldsymbol{C}_{\mathrm{z}i} = |\boldsymbol{C}_{\mathrm{z}i}|\,, \quad \boldsymbol{L} = |\boldsymbol{L}|$$

is true for all $i \in \mathcal{N}$ and $t \geq 0$.

6.3.5 Stability of interconnected continuous state-feedback loops

This section compares the condition (6.40) obtained in the previous section with a stability condition that guarantees the stability of the reference systems

$$\Sigma_{\mathrm{r}i}: \begin{cases} \dot{\boldsymbol{x}}_{\mathrm{r}i}(t) = \bar{\boldsymbol{A}}_i \boldsymbol{x}_{\mathrm{r}i}(t) + \boldsymbol{E}_i \boldsymbol{d}_i(t) + \boldsymbol{E}_{\mathrm{s}i} \boldsymbol{s}_{\mathrm{r}i}(t), & \boldsymbol{x}_{\mathrm{r}i}(0) = \boldsymbol{x}_{0i} \\ \boldsymbol{z}_{\mathrm{r}i}(t) = \boldsymbol{C}_{\mathrm{z}i} \boldsymbol{x}_{\mathrm{r}i}(t) \end{cases} \tag{6.53}$$

with the interconnection

$$\boldsymbol{s}_{\mathrm{r}i}(t) = \sum_{j=1}^{N} \boldsymbol{L}_{ij} \boldsymbol{z}_{\mathrm{r}j}(t). \tag{6.54}$$

The following theorem states a condition for the stability of the interconnected reference systems that is also derived using the comparison principle.

Theorem 6.3 *The stability of the isolated reference systems* (6.31) *implies the stability of the interconnected reference systems* (6.53), (6.54) *if the condition*

$$\lambda_{\mathrm{P}}\left(\int_0^{\infty} \bar{\boldsymbol{G}}_{\mathrm{xs}}(t)\mathrm{d}t\,|\boldsymbol{L}|\,|\boldsymbol{C}_{\mathrm{z}}|\right) < 1 \tag{6.55}$$

is satisfied, where the matrices $\bar{\boldsymbol{G}}_{\mathrm{xs}}(t)$ *and* $\boldsymbol{C}_{\mathrm{z}}$ *are given in* (6.38) *or* (6.39), *respectively.*

Proof. See Appendix B.3. □

A comparison of the relations (6.40), (6.55) shows that both stability conditions are identical. Hence, the stability of the interconnected reference systems (6.53), (6.54) which is proven using condition (6.55) implies the stability of the interconnected event-based state-feedback loops

(6.12a), (6.12b), (6.32), (6.34). In other words, this analysis has highlighted that the event-based implementation of a decentralized state-feedback controller as proposed in Sec. 6.3.1 does not impose more restrictions on the interconnection relation (6.12b) than are claimed for stability of the interconnected reference systems.

6.3.6 Ultimate bound

Based on the previous analysis results, this section derives a bound on the asymptotically stable set $\mathcal{A}$ for the interconnected event-based state-feedback loops (6.12a), (6.12b), (6.32), (6.34). For the following investigation it is assumed that the stability condition (6.40) is satisfied.

Consider the comparison system (6.49) which yields the bound

$$|\boldsymbol{x}(t)| \leq \boldsymbol{G} * \bar{\boldsymbol{F}}(t)\,|\boldsymbol{x}_0| + \boldsymbol{G} * \left(\boldsymbol{e}_\Delta + \bar{\boldsymbol{G}}_{\mathrm{xd}} * |\boldsymbol{d}|\right).$$

on the state $\boldsymbol{x}(t)$ of the event-based overall system. In general, the disturbance $\boldsymbol{d}(t)$ is unknown, but bounded according to Eq. (2.2). Hence,

$$\bar{\boldsymbol{G}}_{\mathrm{xd}} * |\boldsymbol{d}| \leq \int_0^t \bar{\boldsymbol{G}}_{\mathrm{xd}}(t)\mathrm{d}t \cdot \bar{\boldsymbol{d}} \leq \underbrace{\int_0^\infty \bar{\boldsymbol{G}}_{\mathrm{xd}}(t)\mathrm{d}t}_{=:\,\boldsymbol{M}_{\mathrm{xd}}} \cdot\, \bar{\boldsymbol{d}}$$

holds which is substituted in the previous inequality:

$$|\boldsymbol{x}(t)| \leq \boldsymbol{G} * \bar{\boldsymbol{F}}(t)\,|\boldsymbol{x}_0| + \boldsymbol{G} * \left(\boldsymbol{e}_\Delta + \boldsymbol{M}_{\mathrm{xd}}\bar{\boldsymbol{d}}\right) = \boldsymbol{r}_{\mathrm{x}}(t). \tag{6.56}$$

Note that term which depends upon the initial state $\boldsymbol{x}_0$ asymptotically converges to the origin. In contrast to this, the term $(\boldsymbol{e}_\Delta + \boldsymbol{M}_{\mathrm{xd}}\bar{\boldsymbol{d}})$ is constant and non-vanishing. Hence, for $t \to \infty$ the plant state $\boldsymbol{x}(t)$ is bounded by

$$\limsup_{t\to\infty} |\boldsymbol{x}(t)| \leq \lim_{t\to\infty} \boldsymbol{r}_{\mathrm{x}}(t) = \int_0^\infty \boldsymbol{G}(t)\mathrm{d}t \left(\boldsymbol{e}_\Delta + \boldsymbol{M}_{\mathrm{xd}}\bar{\boldsymbol{d}}\right) =: \boldsymbol{b},$$

where the vector $\boldsymbol{b}$ is referred to as the *ultimate bound.* The integral of the impulse response matrix $\boldsymbol{G}(t)$ follows from Eq. (6.52). Thus, the bound $\boldsymbol{b}$ can be reformulated as

$$\boldsymbol{b} = \hat{\boldsymbol{G}}^{-1}\left(\boldsymbol{e}_\Delta + \boldsymbol{M}_{\mathrm{xd}}\bar{\boldsymbol{d}}\right). \tag{6.57}$$

Consequently, the state $\boldsymbol{x}(t)$ of the event-based control loop (6.12a), (6.12b), (6.32), (6.34) is bounded for $t \to \infty$ to the set

$$\mathcal{A} := \{ \boldsymbol{x} \in \mathbb{R}^n \mid |\boldsymbol{x}| \leq \boldsymbol{b} \}. \tag{6.58}$$

The next theorem summarizes the analysis result.

Theorem 6.4 *Consider the interconnected event-based state-feedback loops* (6.12a), (6.12b), (6.32), (6.34) *and assume that the stability condition* (6.40) *is satisfied. Then the overall event-based control system is practically stable with respect to the set $\mathcal{A}$ given in* (6.58) *with the ultimate bound* $\boldsymbol{b}$ *as in* (6.57).

The size of the set $\mathcal{A}$ can be adjusted by an appropriate choice of the event threshold vectors $\bar{\boldsymbol{e}}_i$ for all $i \in \mathcal{N}$. To see this, consider Eq. (6.45) which reads

$$\boldsymbol{e}_{\Delta i} = \int_0^\infty \left| \mathrm{e}^{\bar{\boldsymbol{A}}_i \tau} \boldsymbol{B}_i \boldsymbol{K}_i \right| \mathrm{d}\tau \cdot \bar{\boldsymbol{e}}_i. \tag{6.59}$$

Hence, the elements in the vector $\boldsymbol{e}_\Delta$ linearly depend upon the event threshold vectors $\bar{\boldsymbol{e}}_i$ which means that the set $\mathcal{A}$ can be made smaller by decreasing the elements in $\bar{\boldsymbol{e}}_i$ and vice versa.

When considering the interconnection of the reference systems (6.53) a similar analysis as before leads to the following result on the stability of the interconnected reference systems (6.53), (6.54).

Theorem 6.5 *Consider the interconnected reference systems* (6.53), (6.54) *and assume that the stability condition* (6.55) *is satisfied. Then the overall reference system* (6.53), (6.54) *is practically stable with respect to the set*

$$\mathcal{A}_\mathrm{r} := \{ \boldsymbol{x}_\mathrm{r} \in \mathbb{R}^n \mid |\boldsymbol{x}_\mathrm{r}| \leq \boldsymbol{b}_\mathrm{r} \}$$

with the ultimate bound

$$\boldsymbol{b}_\mathrm{r} = \hat{\boldsymbol{G}}^{-1} \boldsymbol{M}_\mathrm{xd} \bar{\boldsymbol{d}}. \tag{6.60}$$

Proof. See Appendix B.4. □

Consider the bounds $\boldsymbol{b}$ and $\boldsymbol{b}_\mathrm{r}$ given in (6.57) or (6.60), respectively. A comparison of these equations reveals that

$$\boldsymbol{b} = \boldsymbol{b}_\mathrm{r} + \hat{\boldsymbol{G}}^{-1} \boldsymbol{e}_\Delta$$

holds, i.e., the difference between the bounds $\boldsymbol{b}$ and $\boldsymbol{b}_\mathrm{r}$ only depends upon $\boldsymbol{e}_\Delta$ which is a function of the event threshold vectors $\bar{\boldsymbol{e}}_i$ ($i \in \mathcal{N}$). From Eq. (6.59) it can be inferred that this difference can be made arbitrarily small by choosing sufficiently small event thresholds $\bar{e}_i$ for the event generators E_i $(i \in \mathcal{N})$.

6.3.7 Minimum inter-event time

This section derives a bound $\bar{T}_i$ on the minimum inter-event time

$$T_{\min i} := \min_{k_i} \left(t_{k_i+1} - t_{k_i} \right), \quad \forall\, k_i \in \mathbb{N}_0$$

for two consecutive events triggered by the event generator E_i $(i \in \mathcal{N})$. Although the event-based control loops are designed according to the event-based state-feedback approach presented in Ch. 3, the analysis method for the minimum inter-event time given in Theorem 3.2 does not apply to the interconnected event-based state-feedback loops. The following analysis shows that the minimum inter-event time for the interconnected event-based control loops depends upon the coupling input $\boldsymbol{s}_i(t)$ and the main problem to be solved subsequently is to determine a bound on these coupling signals. Naturally, such a bound only exists if the overall control systems is stable and, thus, the stability condition (6.40) is assumed to be fulfilled.

The investigation of the inter-event time that is presented next, rests upon the results on the boundedness of the plant state $\boldsymbol{x}(t)$ that are obtained in the previous section by means of the comparison system (6.56). The main result of this analysis is summarized in the following theorem.

Theorem 6.6 *Consider the event-based state-feedback loops* (6.12a), (6.32), (6.34) *with their interconnections described in* (6.12b). *Assume that the relation* (6.40) *holds. Then, the minimum inter-event time $T_{\min i}$ is bounded from below by $\bar{T}_i$ with*

$$\bar{T}_i := \arg\min_t \left\{ \int_0^t \left| \mathrm{e}^{\boldsymbol{A}_i t} \right| \mathrm{d}t \left(|\boldsymbol{E}_i|\, \bar{\boldsymbol{d}}_{\Delta i} + |\boldsymbol{E}_{\mathrm{s}i}|\, \hat{\boldsymbol{r}}_{\mathrm{s}i} \right) = \bar{e}_i \right\} \tag{6.61}$$

where $\bar{\boldsymbol{d}}_{\Delta i} \geq \left| \boldsymbol{d}_i(t) - \hat{\boldsymbol{d}}_{i,k_i} \right|$ denotes the maximum deviation between the local disturbance $\boldsymbol{d}_i(t)$ and the corresponding estimate $\hat{\boldsymbol{d}}_{i,k_i}$ for all $t \geq 0$ and all $k \in \mathbb{N}_0$ and

$$\hat{\boldsymbol{r}}_{\mathrm{s}} = \begin{pmatrix} \hat{\boldsymbol{r}}_{\mathrm{s}1} \\ \vdots \\ \hat{\boldsymbol{r}}_{\mathrm{s}N} \end{pmatrix} = |\boldsymbol{L}|\, |\boldsymbol{C}_{\mathrm{z}}| \left(\hat{\boldsymbol{G}}^{-1} \left(\sup_{t \geq 0} \bar{\boldsymbol{F}}(t) \mathrm{d}t\, |\boldsymbol{x}_0| \right) + \hat{\boldsymbol{G}}^{-1} \left(\boldsymbol{e}_\Delta + \boldsymbol{M}_{\mathrm{xd}} \bar{\boldsymbol{d}} \right) \right) \tag{6.62}$$

denotes a bound on the maximum magnitude of the coupling input $|\boldsymbol{s}(t)|$ for all $t \geq 0$.

The proof of Theorem 6.6 is presented at the end of this section.

First note that the equation in (6.61) is to be understood to hold element-wise, i.e., $\bar{T}_i$ is the minimum time t for which any element of the vector on the left-hand side equals the corresponding element of the event threshold vector $\bar{\boldsymbol{e}}_i$. In the following the main analysis result is discussed.

Theorem 6.6 shows that the need for a feedback communication from E_i to C_i can have two reasons. First, the disturbance estimate $\hat{\boldsymbol{d}}_{i,k_i}$ deviates from local disturbance $\boldsymbol{d}_i(t)$ and, second, the subsystems Σ_j (that are interconnected with Σ_i) perturb Σ_i via the coupling signal $\boldsymbol{s}_i(t)$. The analysis result reflects the intuition that a larger coupling input $\boldsymbol{s}_i(t)$ yields a shorter inter-event time $T_{\min i}$. Note that Theorem 6.6 gives an explicit bound on the coupling signal $\boldsymbol{s}(t)$, which depends upon the event threshold vectors $\bar{\boldsymbol{e}}_i$ of all event generators E_i ($i \in \mathcal{N}$), the overall disturbance $\boldsymbol{d}(t)$ and the initial state $\boldsymbol{x}_0$ of the plant. Consider the special case that $\boldsymbol{x}_0 = 0$ holds, then Eq. (6.62) reduces to

$$\hat{\boldsymbol{r}}_{\text{s}} = |\boldsymbol{L}|\,|\boldsymbol{C}_{\text{z}}|\,\boldsymbol{b} \geq |\boldsymbol{s}(t)|\,, \quad \forall\, t \geq 0$$

with the ultimate bound $\boldsymbol{b}$ given in (6.57). From the interpretation of the bound $\boldsymbol{b}$ subsequent to Theorem 6.4 it can be inferred that the effect of the coupling input can be manipulated by the choice of all event threshold vectors $\bar{\boldsymbol{e}}_i$.

Remark 6.1 *The knowledge that the effect of the coupling input $\boldsymbol{s}_i(t)$ on subsystem Σ_i can be manipulated to some degree by the choice of the event thresholds $\bar{\boldsymbol{e}}_j$ ($j \in \mathcal{N} \setminus \{i\}$) is used in the approach to event-based state feedback with incomplete state measurement presented in Sec. 5.4 to ensure a desired performance of the overall event-based control system.*

The proof of Theorem 6.6 makes use of the following lemma.

Lemma 6.2 *Consider the interconnected event-based state-feedback loops* (6.12a), (6.12b), (6.32), (6.34). *The minimum inter-event time $T_{\min i}$ for two consecutive events triggered by event generator E_i is bounded from below by*

$$\bar{T}_i := \arg\min_t \left\{ \int_0^t \left|\mathrm{e}^{\boldsymbol{A}_i t}\right| \mathrm{d}t \left(|\boldsymbol{E}_i|\, \bar{\boldsymbol{d}}_{\Delta i} + |\boldsymbol{E}_{\text{s}i}| \cdot \sup_{t \geq 0} |\boldsymbol{s}_i(t)| \right) = \bar{\boldsymbol{e}}_i \right\} \tag{6.63}$$

where $\bar{\boldsymbol{d}}_{\Delta i} \geq \left|\boldsymbol{d}_i(t) - \hat{\boldsymbol{d}}_{i,k_i}\right|$ is the maximum disturbance estimation error for all $t \geq 0$ and all $k \in \mathbb{N}_0$.

Proof. See Appendix B.5. □

The bound $\bar{T}_i$ given in (6.63) results from the analysis that investigates how fast the difference state $|\boldsymbol{x}_{\Delta i}(t)|$ reaches the event threshold $\bar{\boldsymbol{e}}_i$. Now the proof of Theorem 6.6 is presented.

Proof. Consider Eq. (6.63) which presents a bound on the minimum inter-event time in dependence upon the maximum magnitude of the coupling input signal $\boldsymbol{s}_i(t)$. The following aims at deriving a bound on this magnitude for all $t \geq 0$. To this end consider the overall coupling input $\boldsymbol{s}(t)$ which is bounded by

$$|\boldsymbol{s}(t)| \leq |\boldsymbol{L}|\,|\boldsymbol{C}_{\mathrm{z}}|\,|\boldsymbol{x}(t)|\,, \quad \forall\, t \geq 0.$$

Assume that the condition (6.40) holds. Then, by substituting Eq. (6.56) into the last inequality the relation

$$|\boldsymbol{s}(t)| \leq |\boldsymbol{L}|\,|\boldsymbol{C}_{\mathrm{z}}|\,\boldsymbol{G} * \left(\bar{\boldsymbol{F}}(t)\,|\boldsymbol{x}_0| + \boldsymbol{e}_\Delta + \boldsymbol{M}_{\mathrm{xd}}\bar{\boldsymbol{d}}\right)$$

follows. Taking the supremum of both sides yields

$$\begin{aligned}\sup_{t\geq 0}|\boldsymbol{s}(t)| &\leq |\boldsymbol{L}|\,|\boldsymbol{C}_{\mathrm{z}}|\left(\sup_{t\geq 0}\boldsymbol{G} * \bar{\boldsymbol{F}}(t)\,|\boldsymbol{x}_0| + \int_0^\infty \boldsymbol{G}(t)\mathrm{d}t\,\left(\boldsymbol{e}_\Delta + \boldsymbol{M}_{\mathrm{xd}}\bar{\boldsymbol{d}}\right)\right)\\ &\leq |\boldsymbol{L}|\,|\boldsymbol{C}_{\mathrm{z}}|\left(\int_0^\infty \boldsymbol{G}(t)\mathrm{d}t\left(\sup_{t\geq 0}\bar{\boldsymbol{F}}(t)\,|\boldsymbol{x}_0|\right) + \int_0^\infty \boldsymbol{G}(t)\mathrm{d}t\,\left(\boldsymbol{e}_\Delta + \boldsymbol{M}_{\mathrm{xd}}\bar{\boldsymbol{d}}\right)\right).\end{aligned}$$

With Eq. (6.52) the bound

$$\sup_{t\geq 0}|\boldsymbol{s}(t)| \leq |\boldsymbol{L}|\,|\boldsymbol{C}_{\mathrm{z}}|\left(\hat{\boldsymbol{G}}^{-1}\left(\sup_{t\geq 0}\bar{\boldsymbol{F}}(t)\,|\boldsymbol{x}_0|\right) + \hat{\boldsymbol{G}}^{-1}\left(\boldsymbol{e}_\Delta + \boldsymbol{M}_{\mathrm{xd}}\bar{\boldsymbol{d}}\right)\right) =: \hat{\boldsymbol{r}}_{\mathrm{s}}$$

is obtained where

$$\hat{\boldsymbol{r}}_{\mathrm{s}} = \begin{pmatrix}\hat{\boldsymbol{r}}_{\mathrm{s}1}\\ \vdots \\ \hat{\boldsymbol{r}}_{\mathrm{s}N}\end{pmatrix} \geq \begin{pmatrix}|\boldsymbol{s}_1(t)|\\ \vdots \\ |\boldsymbol{s}_N(t)|\end{pmatrix}$$

holds for all $t \geq 0$. Using the bound $\hat{\boldsymbol{r}}_{\mathrm{s}i} \geq \sup_{t\geq 0}|\boldsymbol{s}_i(t)|$ in (6.63) yields Eq. (6.61) and, hence, concludes the proof of Theorem 6.6. □

6.4 Example: Event-based disturbance rejection with local information couplings

This sections studies the disturbance behavior of the event-based state feedback with local information couplings for the example of the thermofluid process introduced in Sec. 2.4. It is considered that the event-based state-feedback controller for each subsystems Σ_1 and Σ_2 of the thermofluid process is designed under the assumption of vanishing interconnections and in this way follows the approach that is investigated in Sec. 6.3.

This example also demonstrates the combination of the proposed event-based state-feedback approach with a nonlinear event-based controller as explained in Sec. 1.3.1. The nonlinear event-based control approach that has been published in [1] is used to drive the state $\boldsymbol{x}(t)$ of the overall system into a target region $\mathcal{T}$ where the decentralized event-based state-feedback controller keeps the state $\boldsymbol{x}(t)$ in spite of disturbances and interconnections between the subsystems.

6.4.1 Specification of the control aim

The state $\boldsymbol{x}(t)$ of the interconnected thermofluid processes (6.12a), (6.12b) with the parameters (A.7), (A.8) should be steered from a given initial state $\boldsymbol{x}_0 \in \mathcal{X}$, with the set $\mathcal{X}$ given in (A.3), into the target region $\mathcal{T} = \mathcal{T}_1 \times \mathcal{T}_2$ with

$$\mathcal{T}_1 = [0.3; 0.36] \text{ m} \times [291.7; 297.7] \text{ K} \tag{6.64a}$$

$$\mathcal{T}_2 = [0.31; 0.37] \text{ m} \times [297.2; 303.2] \text{ K} \tag{6.64b}$$

including the operating point (A.5). Once the state $\boldsymbol{x}(t)$ has entered $\mathcal{T}$ it should be kept in this set by the proposed decentralized event-based state-feedback approach for all time in spite of the influence of disturbances given in (A.4) and interconnections which are set by the valve openings given in (A.8).

6.4.2 Design and analysis of the decentralized event-based state feedback

Design of the event-based controller. The event-based state-feedback controller for each subsystem Σ_i of the thermofluid process is designed according to the event-based state-feedback approach [115], assuming that the subsystems are not interconnected. The feedback gains

$$\boldsymbol{K}_{\text{d1}} = \begin{pmatrix} 10.5 & 0 \\ 0.90 & -0.05 \end{pmatrix}, \quad \boldsymbol{K}_{\text{d2}} = \begin{pmatrix} 11.5 & 0 \\ 1.10 & 0.40 \end{pmatrix} \tag{6.65}$$

are chosen such that the matrices $\boldsymbol{A}_i - \boldsymbol{B}_i \boldsymbol{K}_{\text{d}i}$ $(i = 1, 2)$ are Hurwitz for the parameters given in (A.7). The control input generators C_i determine the control inputs $\boldsymbol{u}_i(t)$ using the model (6.32). The event threshold vectors of the event generators E_i described by (6.34) are set to

$$\bar{\boldsymbol{e}}_1 = \begin{pmatrix} 0.02 \\ 0.4 \end{pmatrix}, \quad \bar{\boldsymbol{e}}_2 = \begin{pmatrix} 0.02 \\ 0.4 \end{pmatrix}, \tag{6.66}$$

which means that E_1 and E_2 trigger an event if either the level or the temperature deviates by $2\ \text{cm}$ or by $0.4\ \text{K}$, respectively, from the corresponding model state.

Analysis of the interconnected event-based control loops. First, the stability of the interconnected event-based state-feedback loops is tested using Algorithm 6.1. The steps of this algorithm are the following:

Step 1: From (A.7) and (A.8) the matrices

$$|\boldsymbol{C}_{\text{z}}| = \operatorname{diag}\left(|\boldsymbol{C}_{\text{z1}}|, |\boldsymbol{C}_{\text{z2}}|\right) = \begin{pmatrix} \boldsymbol{I}_2 & \\ & \boldsymbol{I}_2 \end{pmatrix}$$

and

$$|\boldsymbol{L}| = \begin{pmatrix} & |\boldsymbol{L}_{12}| \\ |\boldsymbol{L}_{21}| & \end{pmatrix} = \begin{pmatrix} & \boldsymbol{I}_2 \\ \boldsymbol{I}_2 & \end{pmatrix}$$

follow.

Step 2: With (6.42), (6.44), (6.47) the matrix

$$\int_0^\infty \bar{\boldsymbol{G}}_{\text{xs}}(t)\text{d}t = \begin{pmatrix} 0.081 & 0 & 0 & 0 \\ 4.175 & 0.517 & 0 & 0 \\ 0 & 0 & 0.082 & 0 \\ 0 & 0 & 2.372 & 0.285 \end{pmatrix}$$

is obtained.

Step 3: With the previously determined matrices the stability test (6.40) yields

$$\lambda_{\mathrm{P}}\left(\int_0^\infty \bar{\boldsymbol{G}}_{\mathrm{xs}}(t)\bar{\boldsymbol{L}}\bar{\boldsymbol{C}}_{\mathrm{z}}\mathrm{d}t\right) = 0.384 < 1.$$

Hence, the interconnected event-based state-feedback loops (6.12a), (6.12b), (6.32), (6.34) with the parameters (A.7), (A.8) are practically stable.

The following investigation shows that the interconnected thermofluid process (6.12a), (6.12b), (A.7), (A.8) with the event-based state feedback (6.32), (6.34) is robustly stable with respect to changes of the coupling strength. Therefore consider the interconnection matrix

$$\boldsymbol{L} = \kappa \begin{pmatrix} & \boldsymbol{L}_{12} \\ \boldsymbol{L}_{21} & \end{pmatrix}$$

with $\kappa \in \mathbb{R}_{+}$. The evaluation of the condition (6.41) yields

$$\kappa < \left(\lambda_{\mathrm{p}}\left(\int_0^\infty \bar{\boldsymbol{G}}_{\mathrm{xs}}(t)\mathrm{d}t\,|\boldsymbol{L}|\,|\boldsymbol{C}_{\mathrm{z}}|\right)\right)^{-1} = \frac{1}{0.384} = 2.604,$$

which means that the overall event-based control system is practically stable for $0 \leq \kappa < 2.604$. According to this result, the stability of the isolated event-based state-feedback loops implies the stability of the overall system for an interconnection that is even more than twice as strong as the one which is applied in the experimental investigations at the real plant.

In the following it is verified that the choice of event threshold vectors (6.66) ensures that the state $\boldsymbol{x}(t)$ is kept within the target set $\mathcal{T}$ specified in (6.64). With Eq. (6.57) the ultimate bound

$$\boldsymbol{b} = (0.018 \;\; 2.56 \;\; 0.027 \;\; 1.36)^{\top}$$

is obtained which implies that the state $\boldsymbol{x}(t)$ is practically stable with respect to the set

$$\mathcal{A} = \{\boldsymbol{x} \mid |\boldsymbol{x} - \bar{\boldsymbol{x}}| \leq \boldsymbol{b}\} \tag{6.67}$$

with the operating point $\bar{\boldsymbol{x}}$ given in (A.5). Hence, the states

$$\boldsymbol{x}_1(t) = \begin{pmatrix} l_{\mathrm{B}}(t) \\ \vartheta_{\mathrm{B}}(t) \end{pmatrix}, \quad \boldsymbol{x}_2(t) = \begin{pmatrix} l_{\mathrm{S}}(t) \\ \vartheta_{\mathrm{S}}(t) \end{pmatrix}$$

are kept within the bounds

$$\begin{aligned} l_{\mathrm{B}}(t) &\in [0.312; 0.348], \qquad \vartheta_{\mathrm{B}}(t) \in [292.1; 297.3] \\ l_{\mathrm{S}}(t) &\in [0.313; 0.367], \qquad \vartheta_{\mathrm{S}}(t) \in [298.8; 301.6] \end{aligned} \tag{6.68}$$

for all $t \geq 0$. A comparison of the bounds (6.68) with the desired target set (6.64) shows that the decentralized event-based controllers with the state-feedback gains (6.65) and the event thresholds (6.66) satisfy the control aim.

The following analysis derives the bounds $\bar{T}_1$ and $\bar{T}_2$ on the minimum inter-event times $T_{\min 1}$ and $T_{\min 2}$, respectively, using the method summarized in Theorem 6.6. Figure 6.8 illustrates the trajectories of the signals

$$\begin{aligned} \boldsymbol{h}_1(t) &= \int_0^t \left| \mathrm{e}^{\boldsymbol{A}_1 t} \right| \mathrm{d}t \left(|\boldsymbol{E}_1| \, \bar{\boldsymbol{d}}_{\Delta 1} + |\boldsymbol{E}_{\mathrm{s}1}| \, \hat{\boldsymbol{r}}_{\mathrm{s}1} \right) \\ \boldsymbol{h}_2(t) &= \int_0^t \left| \mathrm{e}^{\boldsymbol{A}_2 t} \right| \mathrm{d}t \left(|\boldsymbol{E}_2| \, \bar{\boldsymbol{d}}_{\Delta 2} + |\boldsymbol{E}_{\mathrm{s}2}| \, \hat{\boldsymbol{r}}_{\mathrm{s}2} \right) \end{aligned}$$

for the maximum disturbance estimation errors $\bar{\boldsymbol{d}}_{\Delta 1} = 0.2$ and $\bar{\boldsymbol{d}}_{\Delta 2} = 0.5$ and with the maximum magnitude of the coupling input

$$\hat{\boldsymbol{r}}_{\mathrm{s}1} = \begin{pmatrix} 0.027 \\ 5.877 \end{pmatrix}, \qquad \hat{\boldsymbol{r}}_{\mathrm{s}2} = \begin{pmatrix} 0.018 \\ 7.872 \end{pmatrix}$$

determined according to Eq. (6.62). Note that the bound $\bar{T}_i$ is given by that time, where any

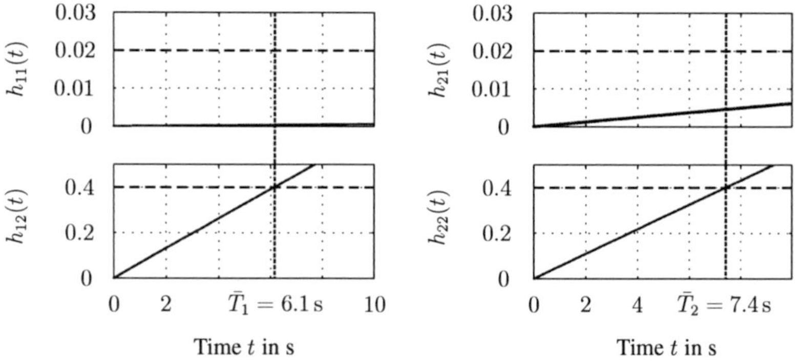

Figure 6.8: Analysis of the minimum inter-event times

element of the vector $\boldsymbol{h}_i(t)$ first equals the corresponding element in the event threshold vector $\bar{\boldsymbol{e}}_i$. The event thresholds are represented in Fig. 6.8 by the horizontal dashed lines. This analysis shows that the minimum inter-event times $T_{\min 1}$ and $T_{\min 2}$ are bounded from below as follows:

$$T_{\min 1} \geq \bar{T}_1 = 6.1\,\mathrm{s}$$
$$T_{\min 2} \geq \bar{T}_2 = 7.4\,\mathrm{s}.$$

The obtained results on the practical stability with respect to the set $\mathcal{A}$ given in (6.68) and on the inter-event times are evaluated in the next section by means of an experiment.

Remark 6.2 *The event generators E_1 and E_2 are implemented as in* (6.34) *with the event condition*

$$|\boldsymbol{x}_{\Delta i}(t)| = |\boldsymbol{x}_i(t) - \boldsymbol{x}_{si}(t)| \geq \bar{\boldsymbol{e}}_i \tag{6.69}$$

for $i = 1, 2$. For the practical realization, the inequality (6.69) *substitutes the event condition in* (6.34) *as due to the implementation on digital hardware the event condition is checked periodically with sampling period $T_s = 0.3\,\mathrm{s}$ which generally only detects the exceeding of the condition. Nevertheless, in relation to the time constants of the process the sampling is fast and, thus, the error that is introduced by the sampling is negligible. A more detailed analysis of event-based control with discrete-time sampling is given in [68].*

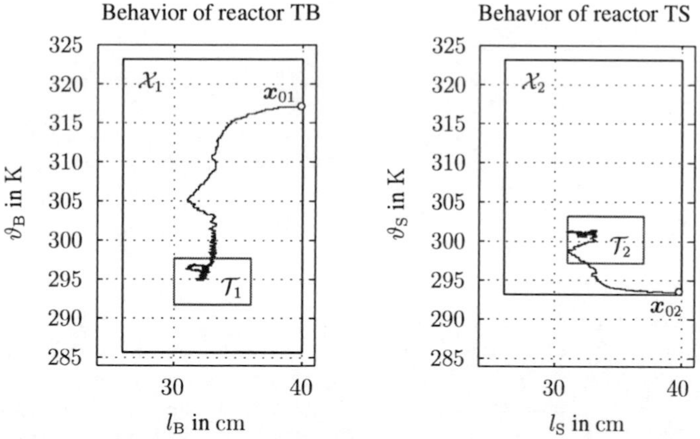

Figure 6.9: Trajectories of the state $\boldsymbol{x}_1(t)$ and $\boldsymbol{x}_2(t)$ in the state-space

6.4.3 Experimental results

The behavior of the continuous flow process with decentralized event-based control is illustrated in Figs. 6.9–6.11. Figure 6.9 gives an overview over the transition of the subsystem states $\boldsymbol{x}_1(t)$ and $\boldsymbol{x}_2(t)$ into the corresponding target regions $\mathcal{T}_1$ and $\mathcal{T}_2$. Once the state $\boldsymbol{x}_1(t)$ or $\boldsymbol{x}_2(t)$ enters $\mathcal{T}_1$ or $\mathcal{T}_2$, respectively, the proposed event-based state-feedback controller is activated which keeps the respective state within the sets. Note that the plant state $\boldsymbol{x}(t)$ is hold in $\mathcal{T}$ despite model uncertainties which occur, since the linearized model (6.12a), (6.12b) with (A.7), (A.8) applied for the event-based controller design, only approximately describes the nonlinear

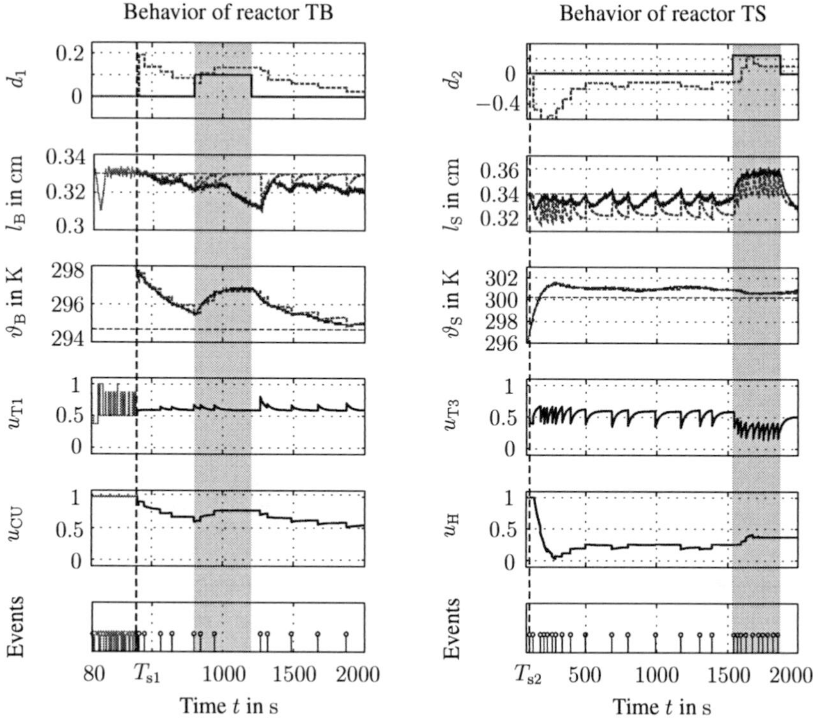

Figure 6.10: Disturbance behavior of the local approach. The behavior of reactor TB and reactor TS is plotted on the left-hand side or right-hand side, respectively. The first row shows the disturbances (solid lines) and the estimated values (dashed lines). The trajectories of the level and temperature are given in the second and third row (solid line: plant state $\boldsymbol{x}(t)$, dashed line: model state $\boldsymbol{x}_s(t)$). The control inputs are illustrated in the next two rows and the event time instants are represented by stems in the bottom figure.

behavior of the plant. Moreover, Figure 6.9 demonstrates the situation where the local control units F_1 and F_2 are activated asynchronously. In reactor TB the target region $\mathcal{T}_1$ is reached within $T_{s1} = 398\,\mathrm{s}$, while in reactor TS the state $\boldsymbol{x}_2(t)$ enters $\mathcal{T}_2$ already after $T_{s2} = 103\,\mathrm{s}$.

The following explanations of the experimental results focus on the situation where the state $\boldsymbol{x}(t)$ is in the set $\mathcal{T}$, describing the behavior of the local approach, i.e., the proposed event-based state feedback with local information couplings. A thorough discussion of the transition of the state $\boldsymbol{x}(t)$ from the initial condition $\boldsymbol{x}_0$ into the target set $\mathcal{T}$ by means of a nonlinear event-based controller can be found in [1].

Figure 6.10 shows the disturbance behavior of the thermofluid process with the proposed event-based state-feedback controller. The areas which are highlighted in gray denote the time intervals in which the disturbances $d_1(t)$ and $d_2(t)$ are active. The time instants $T_{s1} = 398\,\mathrm{s}$ and $T_{s2} = 103\,\mathrm{s}$ at which the local control units F_1 and F_2, respectively, are activated are marked by the vertical dashed lines. The experiments visualize how the feedback communication is adapted to the current system behavior by means of the event-based control. In the time interval $[103, 398]\,\mathrm{s}$, the state $\boldsymbol{x}_2(t)$ is in the target set $\mathcal{T}_2$, whereas $\boldsymbol{x}_1(t)$ is still outside of $\mathcal{T}_1$, which means that reactor TS is considerably affected by reactor TB via the interconnections. In this time interval, 9 events are generated in reactor TS within less than 300 s, while in the time interval $[398, 1550]\,\mathrm{s}$, where the coupling effect is small and the disturbance d_2 is not active, only 7 events are triggered in more than 1150 s. In order to attenuate the disturbance $d_2(t)$ that affects reactor TS in the gray highlighted interval, the feedback communication is induced more often with 10 events being generated. In total, only 13 events within 1602 s are triggered in reactor TB and 26 events within 1897 s are triggered in reactor TS. Compared to a sampled-

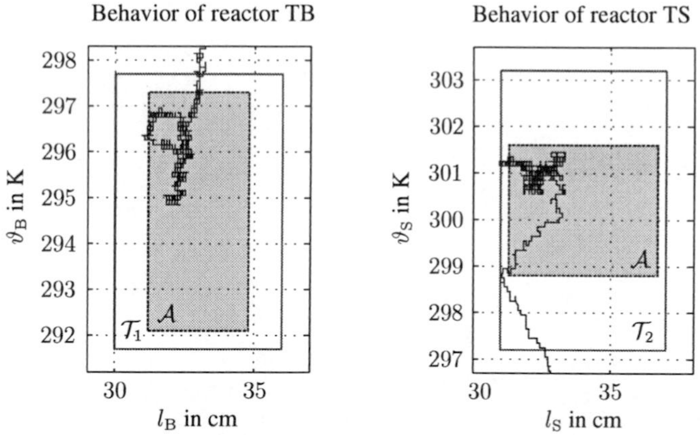

Figure 6.11: Verification of the result on the practical stability with respect to $\mathcal{A}$ given in (6.68)

data control with a sampling period of $h = 10\,\mathrm{s}$ (which is a typical choice for the considered thermofluid process), the feedback communication effort is considerably reduced by the event-based control. The smallest time spans in between consecutive events that are observed in the experiment for both subsystems are

$$T_{\min 1} = 25\,\mathrm{s}, \qquad T_{\min 2} = 10\,\mathrm{s}.$$

The time $T_{\min 1}$ is by a factor of 4 larger than the bound $\bar{T}_1 = 6.1\,\mathrm{s}$, whereas $T_{\min 2}$ is only slightly larger than $\bar{T}_2 = 7.4\,\mathrm{s}$.

Figure 6.11 provides a verification of the bounds (6.68) determined according to the analysis method in Theorem 6.4. The squares drawn with solid lines symbolize the target regions $\mathcal{T}_1$ and $\mathcal{T}_2$ given in (6.64) whereas the set $\mathcal{A}$ determined in (6.67), (6.68) is highlighted in gray. These figures show that in both subsystems Σ_1 and Σ_2, the levels $l_{\mathrm{B}}(t)$ and $l_{\mathrm{S}}(t)$ marginally exceed the calculated bounds. The maximum deviation between the levels and the respective bounds is less than $0.5\,\mathrm{cm}$.

This investigation has shown that that the derived analysis method (summarized in Theorem 6.4 and in Theorem 6.6) yield tight bounds for the considered class of systems. However, the bounds on the asymptotically stable set might not hold in the presence of model uncertainties.

7 Distributed control with event-based communication

This chapter presents an approach to distributed control that combines continuous and event-based state feedback and which aims at the suppression of the disturbance propagation through interconnected systems. First, a new method for the design of distributed state-feedback controllers is proposed and, second, the implementation of the distributed state-feedback law in an event-based manner is presented. At the event times the event-based controllers transmits information to or request information from the neighboring subsystems, which is a new communication pattern that contrasts with one of the event-based control methods that are presented in the previous chapters. The distributed event-based state-feedback approach is tested on the thermofluid process through a simulation and an experiment.

7.1 Event-based control with multicast communication

The topology of many communication networks features the communication from one node to multiple destinations. For an example, consider a networked system where the nodes communicate by radio with a limited range of transmission. In this case, one node can transmit information to its neighboring nodes but cannot distribute the information within the entire network. In network theory, this type of communication is called multicasting [151]. This chapter is dedicated to the investigation of an event-based control method which uses a multicast communication at the event times. This communication scheme is exemplarily illustrated in Fig. 7.1 where the local control unit F_4 transmits information to F_2 and F_3, but not to F_1.

From the perspective of networked control the study of event-based control with multicast communication is motivated by the consideration that a given physically interconnected system should be controlled by a distributed state feedback. A distributed controller might be necessary for example in the following situations:

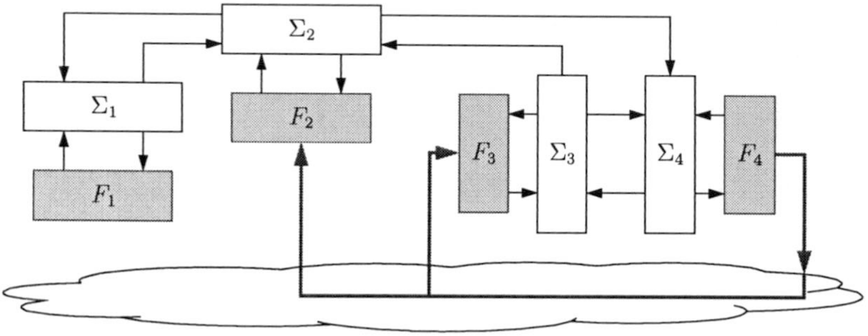

Figure 7.1: Multicast communication scheme

- The overall system cannot be stabilized by a decentralized state feedback (see Chapter 6) and the distributed realization of a centralized controller (see Chapter 5) is undesired because for the vast computational and communication effort.

- The performance (like the suppression of the propagation of the disturbance through the network) of a system with decentralized state feedback should be improved which requires a new controller structure.

This chapter presents a novel method for the implementation of a distributed state-feedback law in an event-based manner.

Section 7.2 introduces a method for the design of a distributed state feedback using the concept of *approximate systems*. This design method is summarized in the Algorithm 7.1. The realization of the obtained distributed state feedback requires continuous communications between several subsystems which would lead to a heavy load of the network and, thus, is undesired. A solution to this problem is presented in Sec. 7.3 in form of a method for the implementation of the distributed state feedback in an event-based fashion where the feedback of the local state information is continuous and the communication between the components of the networked controller occurs at the event times only. The approximation of the behavior of the control system with distributed continuous state feedback by the event-based control is accomplished by a novel triggering mechanism that triggers either the transmission of information to or the request of information from the neighboring subsystems. Section 7.4 presents the results of an experiment that investigates the practical application of the distributed event-based state feedback to the thermofluid process.

The distributed event-based state-feedback approach that is described in this chapter is a generalized version of two methods that have been published in [8, 9]. These works presume the coupling input $\boldsymbol{s}_i(t)$ and the state $\boldsymbol{x}_i(t)$ to be measurable by the control unit F_i for all

$i \in \mathcal{N}$. This assumption is relaxed in this thesis, where each control unit F_i only has access to the subsystem state $\boldsymbol{x}_i(t)$ but not to the coupling input $\boldsymbol{s}_i(t)$.

7.2 A distributed state-feedback design method

This section proposes a new approach to the design of a distributed continuous state-feedback law for physically interconnected systems. In this regard, *distributed* state feedback means that the control input $\boldsymbol{u}_i(t)$ to subsystem Σ_i does not only depend upon the local state $\boldsymbol{x}_i(t)$, but also on the states $\boldsymbol{x}_j(t)$ of the subsystems Σ_j which directly affect Σ_i via the interconnection.

The first part of this section introduces the notion of the *approximate model* and the *extended subsystem*. The underlying idea of the subsequently proposed modeling approach is to obtain a model that does not only describe the behavior of a single subsystem Σ_i, but also includes the interaction of Σ_i with its neighboring subsystems. The extended subsystem model is used for the design of a distributed state-feedback controller.

The main result of this section is summarized in the Algorithm 7.1 for the distributed state-feedback design. The main feature of the proposed design method is that for each subsystem Σ_i the distributed state feedback can be designed without having exact knowledge about the overall system. The stability of the overall control system is tested by means of a condition that is presented in Theorem 7.2.

7.2.1 Approximate model

Regarding the interconnected subsystems (2.3), (2.4)

$$\Sigma_i : \begin{cases} \dot{\boldsymbol{x}}_i(t) = \boldsymbol{A}_i \boldsymbol{x}_i(t) + \boldsymbol{B}_i \boldsymbol{u}_i(t) + \boldsymbol{E}_i \boldsymbol{d}_i(t) + \boldsymbol{E}_{\mathrm{s}i} \boldsymbol{s}_i(t), \quad \boldsymbol{x}_i(0) = \boldsymbol{x}_{0i} \\ \boldsymbol{z}_i(t) = \boldsymbol{C}_{\mathrm{z}i} \boldsymbol{x}_i(t) \end{cases} \tag{7.1a}$$

$$\boldsymbol{s}_i(t) = \sum_{j=1}^{N} \boldsymbol{L}_{ij} \boldsymbol{z}_j(t) \tag{7.1b}$$

the following definitions are made:

Definition 7.1 *Subsystem Σ_j is called* predecessor *of subsystem Σ_i if $\|\boldsymbol{L}_{ij}\| > 0$ holds. In the following*

$$\mathcal{P}_i := \left\{ j \in \mathcal{N} \setminus \{i\} \middle| \, \|\boldsymbol{L}_{ij}\| > 0 \right\}$$

denotes the set of those subsystems Σ_j which directly affect Σ_i through the coupling input $\boldsymbol{s}_i(t)$.

Definition 7.2 *Subsystem Σ_j is called* successor *of subsystem Σ_i if $\|\boldsymbol{L}_{ji}\| > 0$ holds. Hereafter*

$$\mathcal{S}_i := \left\{ j \in \mathcal{N} \setminus \{i\} \middle| \, \|\boldsymbol{L}_{ji}\| > 0 \right\}$$

denotes the set of subsystems Σ_j which the subsystem Σ_i directly affects through the coupling output $\boldsymbol{z}_i(t)$.

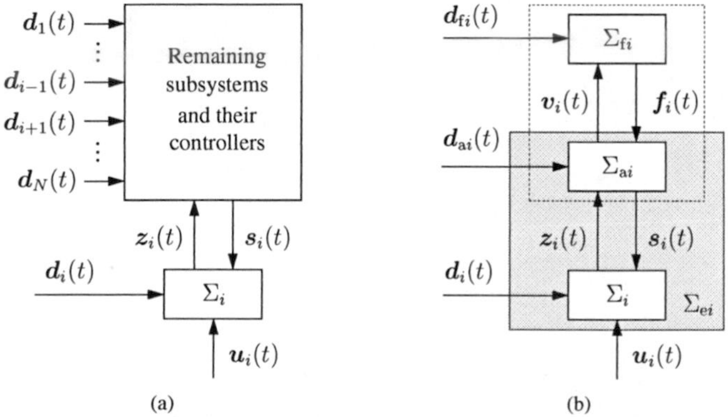

Figure 7.2: Interconnection of subsystem Σ_i and the remaining control loops: (a) general structure; (b) decomposition of the remaining controlled subsystems into approximate model Σ_{ai} and residual model Σ_{fi}.

To understand the notion of the approximate model consider Fig. 7.2. From Eq. (7.1b) follows that the coupling input $\boldsymbol{s}_i(t)$ aggregates the influence of the remaining subsystems together with their controllers on Σ_i (Fig. 7.2(a)). Assume that, from the viewpoint of subsystem Σ_i, the relation between the coupling output $\boldsymbol{z}_i(t)$ and the coupling input $\boldsymbol{s}_i(t)$ is approximately described by the *approximate model*

$$\Sigma_{ai}: \begin{cases} \dot{\boldsymbol{x}}_{ai}(t) = \boldsymbol{A}_{ai}\boldsymbol{x}_{ai}(t) + \boldsymbol{B}_{ai}\boldsymbol{z}_i(t) + \boldsymbol{E}_{ai}\boldsymbol{d}_{ai}(t) + \boldsymbol{F}_{ai}\boldsymbol{f}_i(t), \quad \boldsymbol{x}_{ai}(0) = \boldsymbol{x}_{a0i} \\ \boldsymbol{s}_i(t) = \boldsymbol{C}_{ai}\boldsymbol{x}_{ai}(t) \\ \boldsymbol{v}_i(t) = \boldsymbol{H}_{ai}\boldsymbol{x}_{ai}(t) + \boldsymbol{D}_{ai}\boldsymbol{z}_i(t), \end{cases} \tag{7.2}$$

where $\boldsymbol{x}_{ai} \in \mathbb{R}^{n_{ai}}$, $\boldsymbol{d}_{ai} \in \mathbb{R}^{p_{ai}}$, $\boldsymbol{f}_i \in \mathbb{R}^{\nu_i}$ and $\boldsymbol{v}_i \in \mathbb{R}^{\mu_i}$ denote the state, the disturbance, the residual output and residual input, respectively (Fig. 7.2(b)). The disturbance $\boldsymbol{d}_{ai}(t)$ is assumed to be bounded by

$$|\boldsymbol{d}_{ai}(t)| \leq \bar{\boldsymbol{d}}_{ai}, \quad \forall\, t \geq 0. \tag{7.3}$$

In the following, the state $\boldsymbol{x}_{ai}(t)$ of the approximate model Σ_{ai} is considered to be directly related with the states $\boldsymbol{x}_j(t)$ $(j \in \mathcal{P}_i)$ of the predecessors of Σ_i in the sense that the approximate model Σ_{ai} can be obtained by the linear transformation

$$\boldsymbol{x}_{ai}(t) = \sum_{j \in \mathcal{P}_i} \boldsymbol{T}_{ij}\boldsymbol{x}_j(t), \quad \forall\, t \geq 0 \tag{7.4}$$

with $\boldsymbol{T}_{ij} \in \mathbb{R}^{n_{\mathrm{a}i} \times n_j}$. Moreover, the relation between the subsystem states $\boldsymbol{x}_j(t)$ $(j \in \mathcal{P}_i)$ and the approximate model state $\boldsymbol{x}_{\mathrm{a}i}(t)$ is assumed to be bijective. Hence, for each $j \in \mathcal{P}_i$ there exists a matrix $\boldsymbol{C}_{ji} \in \mathbb{R}^{n_j \times n_{\mathrm{a}i}}$ that satisfies

$$\boldsymbol{x}_j(t) = \boldsymbol{C}_{ji}\boldsymbol{x}_{\mathrm{a}i}(t), \quad \forall\, t \geq 0. \tag{7.5}$$

Note that Eqs. (7.4), (7.5) imply

$$\boldsymbol{C}_{ji}\boldsymbol{T}_{ip} = \begin{cases} \boldsymbol{I}_{n_j}, & \text{if } j = p \\ \boldsymbol{O}, & \text{if } j \neq p. \end{cases}$$

The model $\Sigma_{\mathrm{a}i}$ approximately describes the behavior of the predecessor subsystems Σ_j $(j \in \mathcal{P}_i)$ together with their controllers. This consideration leads to the following assumption.

A 7.1 The matrix $\boldsymbol{A}_{\mathrm{a}i}$ is Hurwitz for all $i \in \mathcal{N}$.

The mismatch between the behavior of the overall control system except subsystem Σ_i and the approximate model (7.2) is expressed by the *residual model*

$$\Sigma_{\mathrm{f}i} : \boldsymbol{f}_i(t) = \boldsymbol{G}_{\mathrm{fd}i} * \boldsymbol{d}_{\mathrm{f}i} + \boldsymbol{G}_{\mathrm{fv}i} * \boldsymbol{v}_i \tag{7.6}$$

where $\boldsymbol{d}_{\mathrm{f}i} \in \mathbb{R}^{p_{\mathrm{f}i}}$ denotes the disturbance that is assumed to be bounded by

$$|\boldsymbol{d}_{\mathrm{f}i}(t)| \leq \bar{\boldsymbol{d}}_{\mathrm{f}i}, \quad \forall\, t \geq 0. \tag{7.7}$$

Note that the approximate model $\Sigma_{\mathrm{a}i}$ together with the residual model $\Sigma_{\mathrm{f}i}$ represents the behavior of the remaining subsystems and their controllers (Fig. 7.2(b)). Hereafter the residual model $\Sigma_{\mathrm{f}i}$ is not assumed to be known exactly but described by some upper bounds $\bar{\boldsymbol{G}}_{\mathrm{fd}i}(t)$ and $\bar{\boldsymbol{G}}_{\mathrm{fv}i}(t)$ which satisfy the relations

$$\bar{\boldsymbol{G}}_{\mathrm{fd}i}(t) \geq |\boldsymbol{G}_{\mathrm{fd}i}(t)|\,, \quad \bar{\boldsymbol{G}}_{\mathrm{fv}i}(t) \geq |\boldsymbol{G}_{\mathrm{fv}i}(t)| \tag{7.8}$$

for all $t \geq 0$.

7.2.2 Extended subsystem model

Consider the subsystem Σ_i augmented with the approximate model $\Sigma_{\mathrm{a}i}$ which yields the *extended subsystem*

$$\Sigma_{\mathrm{e}i} : \begin{cases} \dot{\boldsymbol{x}}_{\mathrm{e}i}(t) = \boldsymbol{A}_{\mathrm{e}i}\boldsymbol{x}_{\mathrm{e}i}(t) + \boldsymbol{B}_{\mathrm{e}i}\boldsymbol{u}_i(t) + \boldsymbol{E}_{\mathrm{e}i}\boldsymbol{d}_{\mathrm{e}i}(t) + \boldsymbol{F}_{\mathrm{e}i}\boldsymbol{f}_i(t), \\ \boldsymbol{x}_{\mathrm{e}i}(0) = \boldsymbol{x}_{\mathrm{e}0i} = \begin{pmatrix} \boldsymbol{x}_{0i}^{\top} & \boldsymbol{x}_{\mathrm{a}0i}^{\top} \end{pmatrix}^{\top} \\ \boldsymbol{v}_i(t) = \boldsymbol{H}_{\mathrm{e}i}\boldsymbol{x}_{\mathrm{e}i}(t) \end{cases} \tag{7.9}$$

with the state $\boldsymbol{x}_{ei} = (\boldsymbol{x}_i^\top \;\; \boldsymbol{x}_{ai}^\top)^\top \in \mathbb{R}^{n_i+n_{ai}}$, the composite disturbance vector $\boldsymbol{d}_{ei} = (\boldsymbol{d}_i^\top \;\; \boldsymbol{d}_{ai}^\top)^\top$ and the matrices

$$\boldsymbol{A}_{ei} = \begin{pmatrix} \boldsymbol{A}_i & \boldsymbol{E}_{si}\boldsymbol{C}_{ai} \\ \boldsymbol{B}_{ai}\boldsymbol{C}_{zi} & \boldsymbol{A}_{ai} \end{pmatrix}, \quad \boldsymbol{B}_{ei} = \begin{pmatrix} \boldsymbol{B}_i \\ \boldsymbol{O} \end{pmatrix}, \quad \boldsymbol{E}_{ei} = \begin{pmatrix} \boldsymbol{E}_i & \boldsymbol{O} \\ \boldsymbol{O} & \boldsymbol{E}_{ai} \end{pmatrix}, \quad \boldsymbol{F}_{ei} = \begin{pmatrix} \boldsymbol{O} \\ \boldsymbol{F}_{ai} \end{pmatrix}, \tag{7.10a}$$

$$\boldsymbol{H}_{ei} = \begin{pmatrix} \boldsymbol{D}_{ai}\boldsymbol{C}_{zi} & \boldsymbol{H}_{ai} \end{pmatrix}. \tag{7.10b}$$

The boundedness of the disturbances $\boldsymbol{d}_i(t)$ and $\boldsymbol{d}_{ai}(t)$ implies the boundedness of $\boldsymbol{d}_{ei}(t)$. With Eqs. (2.5), (7.3), the bound

$$|\boldsymbol{d}_{ei}(t)| \leq \begin{pmatrix} \bar{\boldsymbol{d}}_i^\top & \bar{\boldsymbol{d}}_{ai}^\top \end{pmatrix}^\top =: \bar{\boldsymbol{d}}_{ei}, \quad \forall\, t \geq 0 \tag{7.11}$$

on the composite disturbance $\boldsymbol{d}_{ei}(t)$ follows.

Example 1 *An extended subsystem model for serially interconnected subsystems*

This example illustrates how an extended subsystem model can be determined for subsystem Σ_1 of the overall systems depicted in Fig. 7.3 that consists of three serially interconnected subsystems. Consider that each subsystem Σ_i (for $i = 1, 2, 3$) is described by the state-space model model (7.1a) *and the subsystems are interconnected as indicated in Fig. 7.3 according to the relation* (7.1b) *with*

$$\boldsymbol{L} = \begin{pmatrix} \boldsymbol{O} & \boldsymbol{L}_{12} & \boldsymbol{O} \\ \boldsymbol{L}_{21} & \boldsymbol{O} & \boldsymbol{L}_{23} \\ \boldsymbol{O} & \boldsymbol{L}_{32} & \boldsymbol{O} \end{pmatrix} = \begin{pmatrix} \boldsymbol{O} & \boldsymbol{I} & \boldsymbol{O} \\ \boldsymbol{I} & \boldsymbol{O} & \boldsymbol{I} \\ \boldsymbol{O} & \boldsymbol{I} & \boldsymbol{O} \end{pmatrix}.$$

It is assumed that the subsystems Σ_2 and Σ_3 are controlled by the state feedback

$$\boldsymbol{u}_i(t) = -\boldsymbol{K}_{di}\boldsymbol{x}_i(t), \quad i = 2, 3$$

which results in the controlled subsystems

$$\bar{\Sigma}_i : \begin{cases} \dot{\boldsymbol{x}}_i(t) = \bar{\boldsymbol{A}}_i\boldsymbol{x}_i(t) + \boldsymbol{E}_i\boldsymbol{d}_i(t) + \boldsymbol{E}_{si}\boldsymbol{s}_i(t), & \boldsymbol{x}_i(0) = \boldsymbol{x}_{0i} \\ \boldsymbol{z}_i(t) = \boldsymbol{C}_{zi}\boldsymbol{x}_i(t) \end{cases}$$

for $i = 2, 3$ and where $\bar{\boldsymbol{A}}_i := (\boldsymbol{A}_i - \boldsymbol{B}_i\boldsymbol{K}_{di})$.

In order to determine the approximate model Σ_{a1} first note that the predecessor subsystem of Σ_1 is the subsystem Σ_2: $\mathcal{P}_1 = \{2\}$. According to Eq. (7.4) the approximate model state

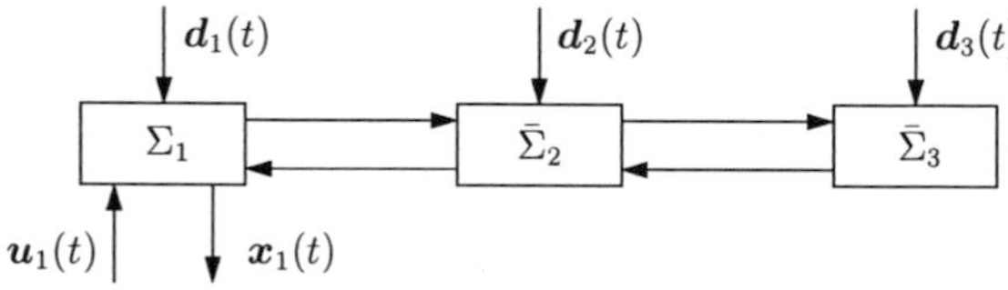

Figure 7.3: Serially interconnetced subsystems

$\boldsymbol{x}_{a1}(t)$ is obtained by means of the transformation

$$\boldsymbol{x}_{a1}(t) = \boldsymbol{x}_2(t)$$

which yields the approximate model

$$\Sigma_{a1}: \begin{cases} \dot{\boldsymbol{x}}_{a1}(t) = \bar{\boldsymbol{A}}_2 \boldsymbol{x}_{a1}(t) + \boldsymbol{E}_{s2}\boldsymbol{z}_1(t) + \boldsymbol{E}_2 \boldsymbol{d}_{a1}(t) + \boldsymbol{E}_{s2}\boldsymbol{f}_1(t), \quad \boldsymbol{x}_{a1}(0) = \boldsymbol{x}_{02} \\ \boldsymbol{s}_1(t) = \boldsymbol{C}_{z2}\boldsymbol{x}_{a1}(t) \\ \boldsymbol{v}_1(t) = \boldsymbol{C}_{z2}\boldsymbol{x}_{a1}(t). \end{cases}$$

Thus, Σ_{a1} is represented by the model (7.2) with

$$\boldsymbol{A}_{a1} = \bar{\boldsymbol{A}}_2, \quad \boldsymbol{B}_{a1} = \boldsymbol{E}_{s2}, \quad \boldsymbol{E}_{a1} = \boldsymbol{E}_2, \quad \boldsymbol{F}_{a1} = \boldsymbol{E}_{s2}, \quad \boldsymbol{C}_{a1} = \boldsymbol{C}_{z2}, \quad \boldsymbol{H}_{a1} = \boldsymbol{C}_{z2}$$

and where $\boldsymbol{d}_{a1}(t) = \boldsymbol{d}_2(t)$ holds. Finally the residual model is given by

$$\Sigma_{f1}: \boldsymbol{f}_1(t) = \boldsymbol{G}_{fd1} * \boldsymbol{d}_{f1} + \boldsymbol{G}_{fv1} * \boldsymbol{v}_1$$

with

$$\boldsymbol{G}_{fd1}(t) = \boldsymbol{C}_{z3} e^{\bar{\boldsymbol{A}}_3 t} \boldsymbol{E}_3, \quad \boldsymbol{G}_{fv1}(t) = \boldsymbol{C}_{z3} e^{\bar{\boldsymbol{A}}_3 t} \boldsymbol{E}_{s3} \tag{7.12}$$

and $\boldsymbol{d}_{f1}(t) = \boldsymbol{d}_3(t)$. Note that the impulse response matrices (7.12) are not required to be known but the bounds

$$\bar{\boldsymbol{G}}_{fd1}(t) \geq \left| \boldsymbol{C}_{z3} e^{\bar{\boldsymbol{A}}_3 t} \boldsymbol{E}_3 \right|, \quad \bar{\boldsymbol{G}}_{fv1}(t) \geq \left| \boldsymbol{C}_{z3} e^{\bar{\boldsymbol{A}}_3 t} \boldsymbol{E}_{s3} \right|$$

are assumed to be given.

The subsystem Σ_1 together with the approximate model Σ_{a1} results in the extended subsystem model

$$\Sigma_{e1}: \begin{cases} \dot{\boldsymbol{x}}_{e1}(t) = \begin{pmatrix} \boldsymbol{A}_1 & \boldsymbol{E}_{s1}\boldsymbol{C}_{z2} \\ \boldsymbol{E}_{s2}\boldsymbol{C}_{z1} & \bar{\boldsymbol{A}}_2 \end{pmatrix} \boldsymbol{x}_{e1}(t) + \begin{pmatrix} \boldsymbol{B}_1 \\ \boldsymbol{O} \end{pmatrix} \boldsymbol{u}_1(t) + \begin{pmatrix} \boldsymbol{E}_1 & \boldsymbol{O} \\ \boldsymbol{O} & \boldsymbol{E}_2 \end{pmatrix} \boldsymbol{d}_{e1}(t) \\ \qquad + \begin{pmatrix} \boldsymbol{O} \\ \boldsymbol{E}_{s2} \end{pmatrix} \boldsymbol{f}_1(t) \\ \boldsymbol{v}_1(t) = \begin{pmatrix} \boldsymbol{O} & \boldsymbol{C}_{z2} \end{pmatrix} \boldsymbol{x}_{e1}(t) \end{cases}$$

with the state $\boldsymbol{x}_{\mathrm{e}1}(t) = \begin{pmatrix}\boldsymbol{x}_1^\top(t) & \boldsymbol{x}_{\mathrm{a}1}^\top(t)\end{pmatrix}^\top = \begin{pmatrix}\boldsymbol{x}_1^\top(t) & \boldsymbol{x}_2^\top(t)\end{pmatrix}^\top$ *and the composite disturbance* $\boldsymbol{d}_{\mathrm{e}1}(t) = \begin{pmatrix}\boldsymbol{d}_1^\top(t) & \boldsymbol{d}_2^\top(t)\end{pmatrix}^\top$. ■

7.2.3 Distributed state-feedback design

The following proposes an approach to the design of a distributed state-feedback controller which is carried out for each subsystem Σ_i of the interconnected subsystems (7.1a), (7.1b) separately. This design approach requires the extended subsystem model (7.9) to be given, in order to determine the state feedback

$$\boldsymbol{u}_i(t) = -\boldsymbol{K}_{\mathrm{e}i}\boldsymbol{x}_{\mathrm{e}i}(t) = -\begin{pmatrix}\boldsymbol{K}_{\mathrm{d}i} & \boldsymbol{K}_{\mathrm{a}i}\end{pmatrix}\begin{pmatrix}\boldsymbol{x}_i(t)\\ \boldsymbol{x}_{\mathrm{a}i}(t)\end{pmatrix}. \tag{7.13}$$

The state-feedback gain $\boldsymbol{K}_{\mathrm{e}i}$ is assumed to be chosen for the extended subsystem (7.9) with $\boldsymbol{f}_i(t) \equiv \boldsymbol{0}$ such that the closed loop system

$$\bar{\Sigma}_{\mathrm{e}i}: \begin{cases}\dot{\boldsymbol{x}}_{\mathrm{e}i}(t) = \bar{\boldsymbol{A}}_{\mathrm{e}i}\boldsymbol{x}_{\mathrm{e}i}(t) + \boldsymbol{E}_{\mathrm{e}i}\boldsymbol{d}_{\mathrm{e}i}(t), & \boldsymbol{x}_{\mathrm{e}i}(0) = \begin{pmatrix}\boldsymbol{x}_{0i}^\top & \boldsymbol{x}_{\mathrm{a}0i}^\top\end{pmatrix}^\top\\ \boldsymbol{v}_i(t) = \boldsymbol{H}_{\mathrm{e}i}\boldsymbol{x}_{\mathrm{e}i}(t)\end{cases} \tag{7.14}$$

has a desired behavior which includes the minimum requirement that the matrix

$$\bar{\boldsymbol{A}}_{\mathrm{e}i} := \boldsymbol{A}_{\mathrm{e}i} - \boldsymbol{B}_{\mathrm{e}i}\boldsymbol{K}_{\mathrm{e}i} = \begin{pmatrix}\boldsymbol{A}_i - \boldsymbol{B}_i\boldsymbol{K}_{\mathrm{d}i} & \boldsymbol{E}_{\mathrm{s}i}\boldsymbol{C}_{\mathrm{a}i} - \boldsymbol{B}_i\boldsymbol{K}_{\mathrm{a}i}\\ \boldsymbol{B}_{\mathrm{a}i}\boldsymbol{C}_{\mathrm{z}i} & \boldsymbol{A}_{\mathrm{a}i}\end{pmatrix} \tag{7.15}$$

is stable. With the transformation (7.4), the control law (7.13) can be rewritten as

$$\boldsymbol{u}_i(t) = -\boldsymbol{K}_{\mathrm{d}i}\boldsymbol{x}_i(t) - \sum_{j\in\mathcal{P}_i}\boldsymbol{K}_{ij}\boldsymbol{x}_j(t) \tag{7.16}$$

with

$$\boldsymbol{K}_{ij} := \boldsymbol{K}_{\mathrm{a}i}\boldsymbol{T}_{ij}.$$

This reveals that (7.13) is a distributed state feedback, where the control input $\boldsymbol{u}_i(t)$ is a function of the local state $\boldsymbol{x}_i(t)$ as well as of the states $\boldsymbol{x}_j(t)$ $(j \in \mathcal{P}_i)$ of the predecessor subsystems of Σ_i.

Note that the proposed design approach neglects the impact of the residual model (7.6) since $\boldsymbol{f}_i(t) \equiv \boldsymbol{0}$ is presumed. Hence, the design of the distributed state feedback only guarantees the stability of the isolated controlled extended subsystems. The next section derives a condition under which the design of the distributed state-feedback controller for the extended subsystem with $\boldsymbol{f}_i(t) \equiv 0$ ensures the stability of the overall control system.

7.2.4 Stability of the overall control system

This section presents a condition to check the stability of the extended subsystem (7.9) with the distributed state feedback (7.16) taking the interconnection to the residual model (7.6), (7.8) into account. The result is then extended to a method for the stability analysis of the overall control system.

The stability analysis method that is presented in this section makes use of the comparison system

$$\boldsymbol{r}_{\mathrm{x}i}(t) = \boldsymbol{V}_{\mathrm{x}0i}(t)\,|\boldsymbol{x}_{\mathrm{e}0i}| + \boldsymbol{V}_{\mathrm{xd}i} * |\boldsymbol{d}_{\mathrm{e}i}| + \boldsymbol{V}_{\mathrm{xf}i} * |\boldsymbol{f}_i| \geq |\boldsymbol{x}_{\mathrm{e}i}(t)| \tag{7.17a}$$

$$\boldsymbol{r}_{\mathrm{v}i}(t) = \boldsymbol{V}_{\mathrm{v}0i}(t)\,|\boldsymbol{x}_{\mathrm{e}0i}| + \boldsymbol{V}_{\mathrm{vd}i} * |\boldsymbol{d}_{\mathrm{e}i}| + \boldsymbol{V}_{\mathrm{vf}i} * |\boldsymbol{f}_i| \geq |\boldsymbol{v}_i(t)|\,, \tag{7.17b}$$

for the extended subsystem (7.9) with the distributed controller (7.16), where

$$\begin{aligned} \boldsymbol{V}_{\mathrm{x}0i}(t) &= \left|\mathrm{e}^{\bar{\boldsymbol{A}}_{\mathrm{e}i}t}\right|, & \boldsymbol{V}_{\mathrm{xd}i}(t) &= \left|\mathrm{e}^{\bar{\boldsymbol{A}}_{\mathrm{e}i}t}\boldsymbol{E}_{\mathrm{e}i}\right|, & \boldsymbol{V}_{\mathrm{xf}i}(t) &= \left|\mathrm{e}^{\bar{\boldsymbol{A}}_{\mathrm{e}i}t}\boldsymbol{F}_{\mathrm{e}i}\right|, \\ \boldsymbol{V}_{\mathrm{v}0i}(t) &= \left|\boldsymbol{H}_{\mathrm{e}i}\mathrm{e}^{\bar{\boldsymbol{A}}_{\mathrm{e}i}t}\right|, & \boldsymbol{V}_{\mathrm{vd}i}(t) &= \left|\boldsymbol{H}_{\mathrm{e}i}\mathrm{e}^{\bar{\boldsymbol{A}}_{\mathrm{e}i}t}\boldsymbol{E}_{\mathrm{e}i}\right|, & \boldsymbol{V}_{\mathrm{vf}i}(t) &= \left|\boldsymbol{H}_{\mathrm{e}i}\mathrm{e}^{\bar{\boldsymbol{A}}_{\mathrm{e}i}t}\boldsymbol{F}_{\mathrm{e}i}\right|. \end{aligned} \tag{7.18}$$

The following lemma gives a condition that can be used to test the stability of the interconnection of the extended subsystem (7.9) with the distributed state feedback (7.16) and the residual model (7.6), (7.8).

Theorem 7.1 *Consider the interconnection of the extended subsystem (7.9) with the distributed state feedback (7.16) together with the residual model $\Sigma_{\mathrm{f}i}$ as defined in (7.6) with the bounds (7.8). The interconnection of these system is practically stable if the condition*

$$\lambda_{\mathrm{P}}\left(\int_0^\infty \bar{\boldsymbol{G}}_{\mathrm{fv}i}(t)\mathrm{d}t \int_0^\infty \boldsymbol{V}_{\mathrm{vf}i}(t)\mathrm{d}t\right) < 1 \tag{7.19}$$

is satisfied.

Proof. See Appendix B.6. □

The relation (7.19) has the following interpretation:

The condition (7.19) can be considered as a small-gain theorem which claims that the extended subsystem (7.9) and the residual model (7.6) are weakly coupled. On the other hand, the condition (7.19) does not impose any restrictions on the interconnection of subsystem Σ_i and its neighboring subsystems Σ_j. These systems do not have to satisfy any small-gain condition but can be even strongly interconnected.

Note that the residual model (7.6) together with the extended subsystem (7.9) with the distributed state feedback (7.16) describe the behavior of the overall control system from the perspective of subsystem Σ_i. Hence, the practical stability of the overall control system is implied by the practical stability of the interconnection of residual model (7.6) and the extended subsystem (7.9) with the controller (7.16) for all $i \in \mathcal{N}$. The following theorem makes the result on the practical stability of the overall control system more precise and determines the set $\mathcal{A}_{ri}$ to which the state $\boldsymbol{x}_i(t)$ is bounded for $t \to \infty$. Hereafter, it is assumed that the condition (7.19) is satisfied which implies that the matrix

$$\int_0^\infty \boldsymbol{G}_i(t)\mathrm{d}t := \left(\boldsymbol{I} - \int_0^\infty \bar{\boldsymbol{G}}_{\mathrm{fv}i}(t)\mathrm{d}t \int_0^\infty \boldsymbol{V}_{\mathrm{vf}i}(t)\mathrm{d}t\right)^{-1}$$

exists and is non-negative.

Theorem 7.2 *Consider the interconnection of the controlled extended subsystem (7.9), (7.16) with the residual model (7.6), (7.8) and assume that the condition (7.19) is satisfied for all $i \in \mathcal{N}$. Then the subsystem Σ_i ($i \in \mathcal{N}$) is practically stable with respect to the set*

$$\mathcal{A}_{ri} := \{\boldsymbol{x}_i \in \mathbb{R}^{n_i} \mid |\boldsymbol{x}_i| \leq \boldsymbol{b}_{ri}\} \tag{7.20}$$

with the ultimate bound

$$\boldsymbol{b}_{\mathrm{r}i} = \begin{pmatrix}\boldsymbol{I}_{n_i} & \boldsymbol{O}\end{pmatrix} \left(\int_0^\infty \boldsymbol{V}_{\mathrm{xd}i}(t)\mathrm{d}t \cdot \bar{\boldsymbol{d}}_{\mathrm{e}i} + \int_0^\infty \boldsymbol{V}_{\mathrm{xf}i}(t)\mathrm{d}t \cdot \bar{\boldsymbol{f}}_i\right), \tag{7.21}$$

where

$$\bar{\boldsymbol{f}}_i = \int_0^\infty \boldsymbol{G}_i(t)\mathrm{d}t \left(\int_0^\infty \bar{\boldsymbol{G}}_{\mathrm{fv}i}(t)\mathrm{d}t \int_0^\infty \boldsymbol{V}_{\mathrm{vd}i}(t)\mathrm{d}t \cdot \bar{\boldsymbol{d}}_{\mathrm{e}i} + \int_0^\infty \bar{\boldsymbol{G}}_{\mathrm{fd}i}(t)\mathrm{d}t \cdot \bar{\boldsymbol{d}}_{\mathrm{f}i}\right)$$

is the bound on the maximum residual output $\boldsymbol{f}_i(t)$ for all $t \geq 0$.

Proof. See Appendix B.7. □

Theorem 7.2 shows that the size of the set $\mathcal{A}_{ri}$ depends upon the disturbance bounds $\bar{\boldsymbol{d}}_{\mathrm{e}i}$ and $\bar{\boldsymbol{d}}_{\mathrm{f}i}$, which together represent a bound on the overall disturbance vector $\boldsymbol{d}(t)$.

7.2.5 Design algorithm

The method for the design of the distributed continuous state-feedback controller is summarized in the following algorithm.

Algorithm 7.1 (Design of a distributed continuous state-feedback controller)

Given: Interconnected system (7.1a), (7.1b).

Do for each $i \in \mathcal{N}$:

1. Identify the set $\mathcal{P}_i$ and determine an approximate model (7.2) and the corresponding residual model (7.6), (7.8). Compose the subsystem model (7.1a) and the approximate model (7.2) to form the extended subsystem model (7.9).
2. Determine the state-feedback gain $\boldsymbol{K}_{\mathrm{e}i}$ as in (7.13) such that the controlled extended subsystem has a desired behavior, implying that the matrix $\bar{\boldsymbol{A}}_{\mathrm{e}i}$ given in (7.15) is Hurwitz.
3. Check whether the condition (7.19) is satisfied. If (7.19) does not hold, stop (the stability of the overall control system cannot be guaranteed).

Result: A state-feedback gain $\boldsymbol{K}_{\mathrm{e}i}$ for each $i \in \mathcal{N}$ which can be implemented in a distributed manner as shown in (7.16). The subsystem Σ_i is practically stable with respect to the set $\mathcal{A}_{\mathrm{r}i}$ given in (7.20) with the ultimate bound (7.21).

Note that the design and the stability analysis of the distributed state feedback can be carried out for each Σ_i $(i \in \mathcal{N})$ without the knowledge of the overall system model. For the design the subsystem model Σ_i and the approximate model $\Sigma_{\mathrm{a}i}$ are required only. In order to analyze the stability of the control system the coarse model information (7.8) of the residual model $\Sigma_{\mathrm{f}i}$ is required.

The crucial point of the design algorithm is in step 3 where the fulfillment of stability condition (7.19) is checked. In case this stability test fails, the following action should be taken.

- Go back to step 2 and determine a new state-feedback gain $\boldsymbol{K}_{\mathrm{e}i}$ such that the matrix

$$\int_0^\infty \boldsymbol{V}_{\mathrm{vf}i}(t)\mathrm{d}t = \int_0^\infty \left| \boldsymbol{H}_{\mathrm{e}i}\mathrm{e}^{(\boldsymbol{A}_{\mathrm{e}i} - \boldsymbol{B}_{\mathrm{e}i}\boldsymbol{K}_{\mathrm{e}i})t}\boldsymbol{F}_{\mathrm{e}i} \right| \mathrm{d}t$$

  is element-wise smaller than for the previous choice of the state-feedback gain. This is generally accomplished by choosing $\boldsymbol{K}_{\mathrm{e}i}$ such that the eigenvalues of the matrix

$$\bar{\boldsymbol{A}}_{\mathrm{e}i} = (\boldsymbol{A}_{\mathrm{e}i} - \boldsymbol{B}_{\mathrm{e}i}\boldsymbol{K}_{\mathrm{e}i})$$

are made smaller compared to the eigenvalues of the same matrix with previously applied $\boldsymbol{K}_{ei}$. Proceed with step 3.

If the stability condition (7.19) still fails after the re-design of the state-feedback gain $\boldsymbol{K}_{ei}$, the next attempt to determine a stabilizig distributed controller according to Algorithm (7.1) should be the following.

- Go back to step 1 and determine a new approximate model $\Sigma_{\mathrm{a}i}$ that describes more accurately the behavior of the neighboring subsystems. Proceed with step 2.

If no approximate model $\Sigma_{\mathrm{a}i}$ can be found for which the condition (7.19) can be satisfied, the proposed design method is not applicable.

The implementation of the distributed state feedback as proposed in (7.16) requires continuous information links from all subsystems Σ_j $(j \in \mathcal{P}_i)$ to subsystem Σ_i. However, continuous communication between several subsystems is in many technical systems neither realizable nor desirable, due to high installation effort and maintenance costs. The following section proposes a method for the implementation of the distributed state-feedback law (7.16) in an event-based fashion where a communication between subsystem Σ_i and its neighboring subsystems Σ_j $(j \in \mathcal{P}_i)$ is induced at the event time instants only.

7.3 Event-based implementation of a distributed state-feedback controller

This section proposes a method for the implementation of a distributed control law of the form (7.16) in an event-based manner where information is exchanged between subsystems at event time instants only. This implementation method obviates the need for a continuous communication among subsystems and it guarantees that the behavior of the event-based control systems approximates the behavior of the control system with distributed continuous state feedback. Following the idea of [115], it will be shown that the deviation between the behavior of these two systems can be made arbitrarily small.

The main results of this section are the following:

- A new approach to the distributed event-based state-feedback control is proposed which can be made to approximate the behavior of the control system with continuous distributed state feedback with arbitrary precision. This feature is guaranteed by means of a new kind of event-based control where two types of events are triggered: Events that are triggered due to the *transmit-condition* induce the transmission of local state information $\boldsymbol{x}_i$ from controller F_i to the controllers F_j of the neighboring subsystems. In addition to this, a second event condition, referred to as *request-condition*, leads to the request of information from the controllers F_j of the neighboring subsystems Σ_j by the controller F_i.

- The deviation between the behavior of the control system with distributed event-based state feedback on the one hand and distributed continuous state feedback on the other hand is bounded (Theorem 7.3).

- The subsystem Σ_i ($i \in \mathcal{N}$) in the distributed event-based state-feedback loop is shown to be practically stable with respect to a set $\mathcal{A}_i$ which is a superset of $\mathcal{A}_{\mathrm{r}i}$ given in (7.20). Moreover, the deviation between the sets $\mathcal{A}_i$ and $\mathcal{A}_{\mathrm{r}i}$ can be made arbitrarily small (Corollary 7.1).

7.3.1 Basic idea

The main idea of the proposed implementation method is as follows: The controller F_i generates a prediction $\tilde{\boldsymbol{x}}_j^i(t)$ of the state $\boldsymbol{x}_j(t)$ for all $j \in \mathcal{P}_i$ which is used to determine the control input $\boldsymbol{u}_i(t)$ according to

$$\boldsymbol{u}_i(t) = -\boldsymbol{K}_{\mathrm{d}i}\boldsymbol{x}_i(t) - \sum_{j \in \mathcal{P}_i} \boldsymbol{K}_{ij}\tilde{\boldsymbol{x}}_j^i(t). \qquad (7.22)$$

This state-feedback law is adapted from (7.16) by replacing the state $\boldsymbol{x}_j(t)$ by its prediction $\tilde{\boldsymbol{x}}_j^i(t)$. Note that the proposed approach to the distributed event-based control allows for different local control units F_i and F_l to determine different predictions $\tilde{\boldsymbol{x}}_j^i(t)$ or $\tilde{\boldsymbol{x}}_j^l(t)$, respectively, of the same state $\boldsymbol{x}_j(t)$ of subsystem Σ_j. These signals are distinguished in the following by means of the superscript.

Following Theorem 4.1, this implementation approach yields a stable overall control system if the difference

$$\left|\boldsymbol{x}_{\Delta j}^i(t)\right| = \left|\boldsymbol{x}_j(t) - \tilde{\boldsymbol{x}}_j^i(t)\right|$$

is bounded for all $t \geq 0$ and all $i, j \in \mathcal{N}$. This difference, however, can neither be monitored by the controller F_i nor by the controller F_j, because the former only knows the prediction $\tilde{\boldsymbol{x}}_j^i(t)$ and the latter has access to the subsystem state $\boldsymbol{x}_j(t)$, only. This problem is solved in the proposed implementation method by a new event triggering mechanism in the controllers F_i. The event generator E_i in F_i generates two different kinds of events: The first one leads to a transmission of the current subsystem state $\boldsymbol{x}_i(t)$ to the controllers F_j of the neighboring subsystems Σ_j, whereas the second one induces the request, denoted by R, of the current state information $\boldsymbol{x}_j(t)$ from all neighboring subsystems Σ_j. In this way, the proposed implementation method introduces a new kind of event-based control, where information is not only send but also requested by the event generators.

7.3.2 Reference system

Assume that for the interconnected subsystems (7.1a), (7.1b) a distributed state-feedback controller is given, which has been designed according to the Algorithm 7.1. The subsystem Σ_i together with the distributed continuous state feedback (7.16) is described by the state-space model

$$\Sigma_{\mathrm{r}i}: \begin{cases} \dot{\boldsymbol{x}}_{\mathrm{r}i}(t) = \bar{\boldsymbol{A}}_i \boldsymbol{x}_{\mathrm{r}i}(t) - \boldsymbol{B}_i \sum\limits_{j \in \mathcal{P}_i} \boldsymbol{K}_{ij} \boldsymbol{x}_{\mathrm{r}j}(t) + \boldsymbol{E}_i \boldsymbol{d}_i(t) + \boldsymbol{E}_{\mathrm{s}i} \boldsymbol{s}_{\mathrm{r}i}(t) \\ \boldsymbol{x}_{\mathrm{r}i}(0) = \boldsymbol{x}_{0i} \\ \boldsymbol{z}_{\mathrm{r}i}(t) = \boldsymbol{C}_{\mathrm{z}i} \boldsymbol{x}_{\mathrm{r}i}(t) \end{cases} \tag{7.23}$$

where $\bar{\boldsymbol{A}}_i := (\boldsymbol{A}_i - \boldsymbol{B}_i \boldsymbol{K}_{\mathrm{d}i})$. Hereafter, $\Sigma_{\mathrm{r}i}$ is referred to as the reference subsystem. The signals are indicated with r to distinguish them from the corresponding signals in the event-based control system that is investigated later. The interconnection of the reference subsystems

Σ_{ri} is given by

$$s_{ri}(t) = \sum_{j=1}^{N} L_{ij} z_{rj}(t) \tag{7.24}$$

according to Eq. (7.1b). Equations (7.23), (7.24) yield the overall reference system

$$\Sigma_r: \quad \dot{x}_r(t) = \underbrace{\left(A - B\left(K_d + \bar{K}\right)\right)}_{:= \bar{A}} x_r(t) + Ed(t), \quad x_r(0) = x_0, \tag{7.25}$$

with the matrices A, B, E given in (2.6) and the state-feedback gains

$$K_d = \mathrm{diag}\left(K_{d1}, \ldots, K_{dN}\right) \tag{7.26}$$

and

$$\bar{K} = \begin{pmatrix} O & K_{12} & \ldots & K_{1N} \\ K_{21} & O & \ldots & K_{2N} \\ \vdots & \vdots & \ddots & \vdots \\ K_{N1} & K_{N2} & \ldots & O \end{pmatrix} \tag{7.27}$$

where $K_{ij} = O$ holds if $j \notin \mathcal{P}_i$. The overall control system (7.25) is stable since the distributed state-feedback gain (7.26), (7.27) is the result of Algorithm 7.1 which includes the validation of the stability condition (7.19). Consequently, the subsystems (7.23) are practically stable with respect to the set $\mathcal{A}_{ri}$ given in (7.20), (7.21).

The aim of the event-based controller that is to be designed subsequently is to approximate the behavior of the reference system (7.23), (7.24) with adjustable precision.

7.3.3 Information transmissions and requests

The event-based control approach that is investigated in this section works with a triggering mechanism that distinguishes between two kind of events which lead to either the transmission of local state information or the request of information from neighboring subsystems. The event conditions, based on which these events are triggered, are called *transmit-condition* and *request-condition*. This event triggering scheme contrast with the one of event-based control methods that are proposed in the previous chapters and, thus, necessitates an extended notion of the triggering time instants.

According to the subsequently presented event-based control approach, the local control unit F_i sends and receives information. Both the sending and the reception of information can occur in two situations which are explained next:

- The local control unit F_i transmits the current state $\boldsymbol{x}_i(t_{k_i})$ if at time t_{k_i} its transmit-condition if fulfilled. On the other hand, F_i sends the state $\boldsymbol{x}_i(t_{r_j(i)})$ at time $t_{r_j(i)}$ when the controller F_j requests this information from F_i.

- If at time $t_{r_i(j)}$ the request-condition of the local control unit F_i is satisfied, it requests (and consequently receives) the information $\boldsymbol{x}_j(t_{r_i(j)})$ from the controller F_j. The controller F_i also receives information if the controller F_j decides at time t_{k_j} to transmit the state $\boldsymbol{x}_j(t_{k_j})$ to the F_i.

In the following, $r_i(j)$ denotes the counter for the events at which the controller F_i requests information from the controller F_j.

> From the viewpoint of the controller F_i the transmission and the reception of information is either induced by its own triggering conditions (at the times t_{k_i} or $t_{r_i(j)}$) or enforced by the neighboring controllers (at the times $t_{r_j(i)}$ or t_{k_j}).

7.3.4 Networked controller

The networked controller consists of the local components F_i and the communication network as shown in Fig. 4.1. Each control unit F_i includes a control input generator C_i that generates the control input $\boldsymbol{u}_i(t)$ and an event generator E_i that determines the time instants t_{k_i} and $t_{r_i(j)}$ at which the current local state $\boldsymbol{x}_i(t_{k_i})$ is communicated to the controllers F_j of the successor subsystems Σ_j $(j \in \mathcal{S}_i)$ or at which the current state $\boldsymbol{x}_j(t_{r_i(j)})$ is requested from the controllers F_j of the predecessor subsystems Σ_j $(j \in \mathcal{P}_i)$, respectively. Figure 7.4 depicts the structure of the local control unit F_i. R denotes a message that initiates the request of information and $t_\star$ represents different event times which are specified later. The components C_i and E_i are explained in the following.

Control input generator C_i. The control input generator C_i determines the control input $\boldsymbol{u}_i(t)$ for subsystem Σ_i using the model

$$C_i : \begin{cases} \tilde{\Sigma}_{\mathrm{ai}} : \begin{cases} \frac{\mathrm{d}}{\mathrm{d}t}\tilde{\boldsymbol{x}}_{\mathrm{ai}}(t) = \boldsymbol{A}_{\mathrm{ai}}\tilde{\boldsymbol{x}}_{\mathrm{ai}}(t) + \boldsymbol{B}_{\mathrm{ai}}\boldsymbol{z}_i(t) \\ \tilde{\boldsymbol{x}}_{\mathrm{ai}}(t_\star^+) = \sum\limits_{p \in \mathcal{P}_i \setminus \{j\}} \boldsymbol{T}_{ip}\boldsymbol{C}_{pi}\tilde{\boldsymbol{x}}_{\mathrm{ai}}(t_\star) + \boldsymbol{T}_{ij}\boldsymbol{x}_j(t_\star), \quad j \in \mathcal{P}_i \end{cases} \\ \boldsymbol{u}_i(t) = -\boldsymbol{K}_i\boldsymbol{x}_i(t) - \boldsymbol{K}_{\mathrm{ai}}\tilde{\boldsymbol{x}}_{\mathrm{ai}}(t). \end{cases} \tag{7.28}$$

The structure of this generator is illustrated in Fig. 7.5. The state $\tilde{\boldsymbol{x}}_{\mathrm{ai}}(t)$ is a prediction of the actual approximate model state $\boldsymbol{x}_{\mathrm{ai}}(t)$ that is generated by means of the model $\tilde{\Sigma}_{\mathrm{ai}}$. Note that

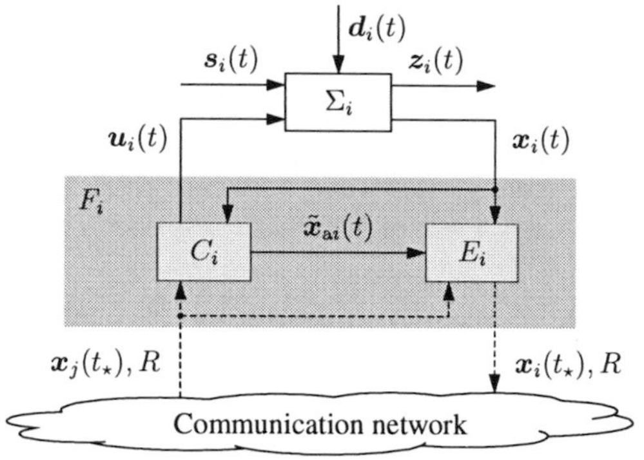

Figure 7.4: Component F_i of the networked controller

$\tilde{\Sigma}_{ai}$ differs from the approximate model Σ_{ai} in that the disturbance $\boldsymbol{d}_{ai}(t)$ and the residual output $\boldsymbol{f}_i(t)$ are omitted, since these signals are unknown to the local controller F_i. The generation of the control input $\boldsymbol{u}_i(t)$ is equivalent with Eq. (7.22), since

$$\boldsymbol{K}_{ai}\tilde{\boldsymbol{x}}_{ai}(t) = \sum_{j\in\mathcal{P}_i} \boldsymbol{K}_{ai}\boldsymbol{T}_{ij}\tilde{\boldsymbol{x}}_j^i(t) = \sum_{j\in\mathcal{P}_i} \boldsymbol{K}_{ij}\tilde{\boldsymbol{x}}_j^i(t).$$

The state $\tilde{\boldsymbol{x}}_{ai}$ is reinitialized at the event times $t_\star$, where the $\star$-symbol is a placeholder either for k_j or for $r_i(j)$. That is, $\tilde{\boldsymbol{x}}_{ai}$ is reinitialized whenever F_i receives current state information $\boldsymbol{x}_j(t_\star)$ from F_j ($j \in \mathcal{P}_i$) which occurs in two situations:

- At time $t = t_{k_j}$, the event generator E_j decides to transmit the current state $\boldsymbol{x}_j(t_{k_j})$ to all controllers F_i of the successor subsystems Σ_j, i.e., to all F_i with $i \in \mathcal{S}_j$.
- At time $t = t_{r_i(j)}$, E_i requests current state information from the controller F_j of the predecessor subsystem Σ_j, i.e., from F_j with $j \in \mathcal{P}_i$.

The two event times t_{k_j} and $t_{r_i(j)}$ are determined based on conditions that are defined in the next paragraph.

Event generator E_i. The task of the event generator E_i is to bound the deviation

$$\left|\boldsymbol{x}_{\Delta j}^i(t)\right| = \left|\boldsymbol{x}_j(t) - \boldsymbol{C}_{ji}\tilde{\boldsymbol{x}}_{ai}(t)\right| = \left|\boldsymbol{x}_j(t) - \tilde{\boldsymbol{x}}_j^i(t)\right| \qquad (7.29)$$

between the actual subsystem state $\boldsymbol{x}_j(t)$ and the prediction $\boldsymbol{C}_{ji}\tilde{\boldsymbol{x}}_{ai}(t) = \tilde{\boldsymbol{x}}_j^i(t)$, applied for the control input generation (7.28). However, the difference (7.29) cannot be monitored by any of the event generators E_i or E_j, which exclusively have access to the prediction $\tilde{\boldsymbol{x}}_{ai}(t)$

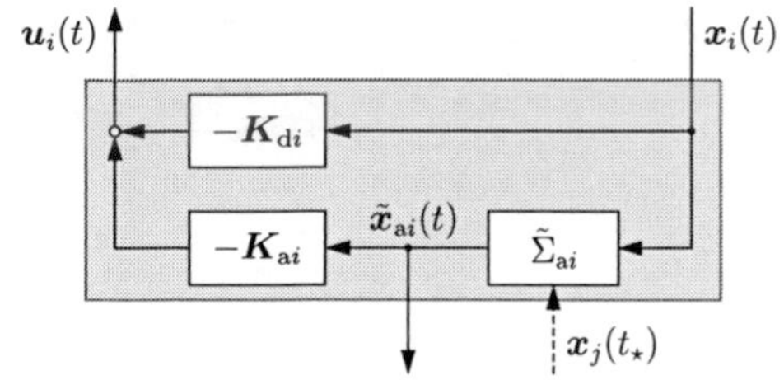

Figure 7.5: Control input generator C_i of the control unit F_i

or to the subsystem state $x_j(t)$, respectively. This consideration gives rise to the idea that the boundedness of (7.29) can be accomplished by a cooperation of the event generators E_i and E_j which is explained in the following.

Consider that both E_i and E_j include the model

$$\Sigma_{cj}: \quad \dot{x}_{cj}(t) = \bar{A}_j x_{cj}(t), \quad x_{cj}(t_\star^+) = x_j(t_\star) \qquad (\text{if } t_\star = t_{k_j} \text{ or } t_\star = t_{r_i(j)}) \tag{7.30}$$

with the matrix $\bar{A}_j = (A_j - B_j K_{dj})$. The state $x_{cj}(t)$ is used in both event generators E_i and E_j as a comparison signal as follows:

- Event generator E_j triggers an event at time t_{k_j} whenever the *transmit-condition*

$$|x_j(t) - x_{cj}(t)| = \alpha_j \bar{e}_j \tag{7.31}$$

  is satisfied, where $\bar{e}_j \in \mathbb{R}_+^{n_j}$ denotes the event threshold vector and $\alpha_j \in (0,1)$ is a weighting factor.

- The event generator E_i triggers an event at time $t_{r_i(j)}$ which induces the request of the state $x_j(t_{r_i(j)})$ from F_j whenever the *request-condition*

$$|C_{ji}\tilde{x}_{ai}(t) - x_{cj}(t)| = (1-\alpha_j)\bar{e}_j, \quad j \in \mathcal{P}_i \tag{7.32}$$

  is fulfilled, with the same event threshold vector $\bar{e}_j \in \mathbb{R}_+^{n_j}$ as in (7.31). That means, at time $t_{r_i(j)}$ the event generator E_i sends a request R to the controller F_j in order to call this controller to transmit the current state $x_j(t_{r_i(j)})$.

The event generator E_j of the controller F_j responses to the triggering of the events caused by the transmit-condition (7.31) or by the request-condition (7.32) with the same action. It sends the current state $x_j(t_\star)$ (with $t_\star = t_{k_j}$ or $t_\star = t_{r_i(j)}$ if the transmit-condition (7.31) or the request-condition (7.32), respectively, has led to the event triggering) to all F_p with $p \in \mathcal{S}_j$

(that is to all controllers of the successor subsystems Σ_p of subsystem Σ_j, which includes F_i). The received information is then used to reinitialize the state of the models (7.28) in all C_p with $p \in \mathcal{S}_j$, as well as the state of the models (7.30) in E_j and in all E_p with $p \in \mathcal{S}_j$. Due to the event triggering and the state resets the relations

$$|\boldsymbol{x}_i(t) - \boldsymbol{x}_{\mathrm{c}i}(t)| \leq \alpha_i \bar{\boldsymbol{e}}_i \tag{7.33a}$$

$$|\boldsymbol{C}_{ji}\tilde{\boldsymbol{x}}_{\mathrm{a}i}(t) - \boldsymbol{x}_{\mathrm{c}j}(t)| \leq (1-\alpha_j)\bar{\boldsymbol{e}}_j \tag{7.33b}$$

hold for all $i \in \mathcal{N}$ and all $j \in \mathcal{P}_i$. By virtue of (7.33) the state difference (7.29) is bounded by

$$\begin{aligned} \left|\boldsymbol{x}^i_{\Delta j}(t)\right| &= |\boldsymbol{x}_j(t) - \boldsymbol{x}_{\mathrm{c}j}(t) + \boldsymbol{x}_{\mathrm{c}j}(t) - \boldsymbol{C}_{ji}\tilde{\boldsymbol{x}}_{\mathrm{a}i}(t)| \\ &\leq |\boldsymbol{x}_j(t) - \boldsymbol{x}_{\mathrm{c}j}(t)| + |\boldsymbol{C}_{ji}\tilde{\boldsymbol{x}}_{\mathrm{a}i}(t) - \boldsymbol{x}_{\mathrm{c}j}(t)| \leq \alpha_j\bar{\boldsymbol{e}}_j + (1-\alpha_j)\bar{\boldsymbol{e}}_j = \bar{\boldsymbol{e}}_j. \end{aligned} \tag{7.34}$$

In summary, the event generator E_i is represented by

$$E_i : \begin{cases} \Sigma_{\mathrm{c}j} \text{ as described in (7.30), for all } j \in \mathcal{P}_i \cup \{i\} \\ t_{k_i+1} := \inf\left\{t > \hat{t}_{\mathrm{Tx}i} \;\middle|\; |\boldsymbol{x}_i(t) - \boldsymbol{x}_{\mathrm{c}i}(t)| = \alpha_i\bar{\boldsymbol{e}}_i\right\} \\ t_{r_i(j)+1} := \inf\left\{t > \hat{t}_{\mathrm{Rx}i(j)} \;\middle|\; |\boldsymbol{C}_{ji}\tilde{\boldsymbol{x}}_{\mathrm{a}i}(t) - \boldsymbol{x}_{\mathrm{c}j}(t)| = (1-\alpha_j)\bar{\boldsymbol{e}}_j\right\}, \quad \forall\, j \in \mathcal{P}_i. \end{cases} \tag{7.35}$$

In (7.35) the time

$$\hat{t}_{\mathrm{Tx}i} := \max\left\{t_{k_i},\ \max_{j \in \mathcal{S}_i}\left\{t_{r_j(i)}\right\}\right\}$$

denotes the last time at which E_i has transmitted the state $\boldsymbol{x}_i$ to the controllers F_j of the suc-

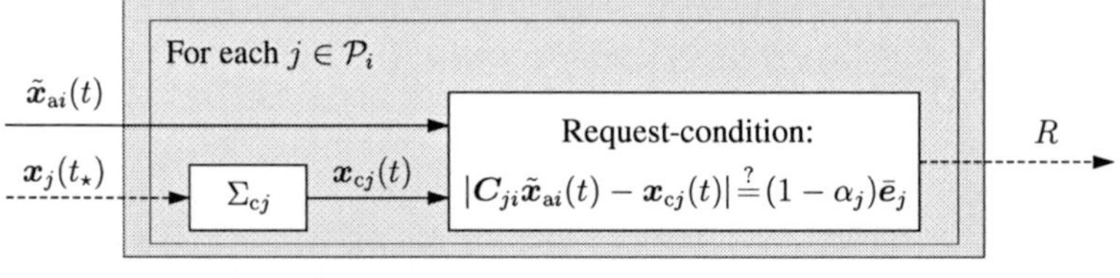

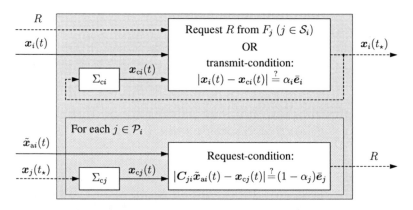

Figure 7.6: Event generator E_i of the control unit F_i

cessor subsystems Σ_j, caused by either the violation of the transmit-condition or the request of information from one of the controllers F_j with $j \in \mathcal{S}_i$. Accordingly, the time

$$\hat{t}_{\mathrm{Rx}i(j)} := \max\left\{t_{r_i(j)}, t_{k_j}\right\}$$

denotes the last time at which E_i has received information about the state $\boldsymbol{x}_j$ either by request or due to the triggering of the transmit-condition in E_j.

Figure 7.6 illustrates the structure of the event generator E_i. Note that the logic that determines the time instants at which the current state $\boldsymbol{x}_i(t_\star)$ (with $t_\star = t_{k_i}$ or $t_\star = t_{r_j}$) is send to the successor subsystems is decoupled from the logic which decides when to request state information from the predecessor subsystems. The latter is represented in the gray block for the example of requesting information from some subsystem Σ_j $(j \in \mathcal{P}_i)$. Note that this logic, highlighted in the gray block, is implemented in E_i for each $j \in \mathcal{P}_i$.

The communication policy leads to a topology for the information exchange that can be characterized as follows:

> The event generator E_i of the controller F_i transmits the state information $\boldsymbol{x}_i$ only to the controllers F_j of the successor subsystems Σ_j $(\in \mathcal{S}_i)$ of Σ_i, whereas F_i receives the state information $\boldsymbol{x}_j$ only from the controllers F_j of the predecessor subsystems Σ_j $(j \in \mathcal{P}_i)$ of Σ_i.

7.3.5 Discussion of the event conditions

This section investigates the triggering conditions (7.31), (7.32) and shows which signals directly affect the triggering of the respective event.

Transmit-condition. Consider the transmit-condition (7.31) and observe that the difference

$$\boldsymbol{\delta}_{\mathrm{c}i}(t) := \boldsymbol{x}_i(t) - \boldsymbol{x}_{\mathrm{c}i}(t)$$

evolves for $t \geq \hat{t}_{\mathrm{Tx}i}$ according to the state-space model

$$\tfrac{\mathrm{d}}{\mathrm{d}t}\boldsymbol{\delta}_{\mathrm{c}i}(t) = \bar{\boldsymbol{A}}_i\boldsymbol{\delta}_{\mathrm{c}i}(t) - \sum_{j\in\mathcal{P}_i} \boldsymbol{B}_i\boldsymbol{K}_{ij}\tilde{\boldsymbol{x}}_j^i(t) + \boldsymbol{E}_i\boldsymbol{d}_i(t) + \boldsymbol{E}_{\mathrm{s}i}\boldsymbol{s}_i(t), \quad \boldsymbol{\delta}_{\mathrm{c}i}(\hat{t}_{\mathrm{Tx}i}) = 0. \tag{7.36}$$

Hence, the triggering of the transmit-condition in E_i is influenced by the predictions $\tilde{\boldsymbol{x}}_j^i(t)$ used for the control input generation, the disturbance $\boldsymbol{d}_i(t)$ and the coupling input signal $\boldsymbol{s}_i(t)$.

Note that $\boldsymbol{\delta}_{\mathrm{c}i}(\hat{t}_{\mathrm{Tx}i}) = 0$ holds due to the reset of the state $\boldsymbol{x}_{\mathrm{c}i}(t)$ whenever E_i transmits the current subsystem state $\boldsymbol{x}_i(t)$, as stated in (7.30). Given that the predictions $\tilde{\boldsymbol{x}}_j^i(t)$ for all $j \in \mathcal{P}_i$, the disturbance $\boldsymbol{d}_i(t)$ and the coupling input signal $\boldsymbol{s}_i(t)$ are bounded, the difference $|\boldsymbol{\delta}_{\mathrm{c}i}(t)|$ cannot attain the threshold $\alpha_i\bar{\boldsymbol{e}}_i$ infinitely fast. This implies that there is a minimum time in between the triggering of two consecutive events by means of the transmit-condition (7.31).

Proposition 7.1 *The minimum inter-event time for consecutive events that are triggered due to the transmit-condition* (7.31) *is bounded from below by a positive constant.*

Request-condition. In order to investigate the difference

$$\boldsymbol{\delta}^i_{\mathrm{c}j}(t) := \boldsymbol{C}_{ji}\tilde{\boldsymbol{x}}_{\mathrm{a}i}(t) - \boldsymbol{x}_{\mathrm{c}j}(t) \tag{7.37}$$

that is monitored in the request-condition (7.32) consider the state-space model

$$\tfrac{\mathrm{d}}{\mathrm{d}t}\boldsymbol{C}_{ji}\tilde{\boldsymbol{x}}_{\mathrm{a}i}(t) = \boldsymbol{C}_{ji}\boldsymbol{A}_{\mathrm{a}i}\tilde{\boldsymbol{x}}_{\mathrm{a}i}(t) + \boldsymbol{C}_{ji}\boldsymbol{B}_{\mathrm{a}i}\boldsymbol{C}_{\mathrm{z}i}\boldsymbol{x}_i(t), \quad \boldsymbol{C}_{ji}\tilde{\boldsymbol{x}}_{\mathrm{a}i}(\hat{t}_{\mathrm{Rx}i(j)}) = \boldsymbol{x}_j(\hat{t}_{\mathrm{Rx}i(j)})$$

which follows from (7.28). With Eq. (7.4) this model can be restated as

$$\begin{aligned}\tfrac{\mathrm{d}}{\mathrm{d}t}\boldsymbol{C}_{ji}\tilde{\boldsymbol{x}}_{\mathrm{a}i}(t) &= \bar{\boldsymbol{A}}_j\tilde{\boldsymbol{x}}^i_j(t) + \sum_{p\in\mathcal{P}_i\setminus\{j\}} \boldsymbol{C}_{ji}\boldsymbol{A}_{\mathrm{a}i}\boldsymbol{T}_{ip}\tilde{\boldsymbol{x}}^i_p(t) + \boldsymbol{C}_{ji}\boldsymbol{B}_{\mathrm{a}i}\boldsymbol{C}_{\mathrm{z}i}\boldsymbol{x}_i(t)\\ \boldsymbol{C}_{ji}\tilde{\boldsymbol{x}}_{\mathrm{a}i}(\hat{t}_{\mathrm{Rx}i(j)}) &= \boldsymbol{x}_j(\hat{t}_{\mathrm{Rx}i(j)})\end{aligned}$$

where the assumption is made that the matrices $\boldsymbol{T}_{ij}$ and $\boldsymbol{C}_{ji}$ are chosen such that

$$\bar{\boldsymbol{A}}_j = \boldsymbol{C}_{ji}\boldsymbol{A}_{\mathrm{a}i}\boldsymbol{T}_{ij}$$

holds. This model together with Eq. (7.30) yield the state-space model

$$\tfrac{\mathrm{d}}{\mathrm{d}t}\boldsymbol{\delta}^i_{\mathrm{c}j}(t) = \bar{\boldsymbol{A}}_j\boldsymbol{\delta}^i_{\mathrm{c}j}(t) + \sum_{p\in\mathcal{P}_i\setminus\{j\}} \boldsymbol{C}_{ji}\boldsymbol{A}_{\mathrm{a}i}\boldsymbol{T}_{ip}\tilde{\boldsymbol{x}}^i_p(t) + \boldsymbol{C}_{ji}\boldsymbol{B}_{\mathrm{a}i}\boldsymbol{C}_{\mathrm{z}i}\boldsymbol{x}_i(t), \quad \boldsymbol{\delta}^i_{\mathrm{c}j}(\hat{t}_{\mathrm{Rx}i(j)}) = \boldsymbol{0} \tag{7.38}$$

that describes the behavior of the difference (7.37) for $t \geq \hat{t}_{\mathrm{Rx}i(j)}$. This shows that the request-condition is directly driven by the predictions $\tilde{\boldsymbol{x}}^i_j(t)$ and the subsystem state $\boldsymbol{x}_i(t)$, but neither the disturbance $\boldsymbol{d}_i(t)$ nor the coupling input $\boldsymbol{s}_i(t)$ have direct impact on the request condition.

For the existence of a minimum inter-event time for two consecutive events triggered by means of the request-condition (7.32) the same argumentation applies as for the transmit-condition. The time in between two consecutive events triggered by the request-condition (7.32) must be strictly positive, since $\boldsymbol{\delta}^i_{cj}(\hat{t}_{\mathrm{Rx}i(j)}) = \boldsymbol{0}$ holds and the predictions $\tilde{\boldsymbol{x}}^i_p(t)$ as well as the subsystem state $\boldsymbol{x}_i(t)$ are bounded. Hence, $\left|\boldsymbol{\delta}^i_{\mathrm{c}j}(t)\right|$ cannot grow infinitely fast to the threshold $(1-\alpha_j)\bar{\boldsymbol{e}}_j$.

Proposition 7.2 *The minimum inter-event time for consecutive events that are triggered due to the request-condition (7.32) is bounded from below by a positive constant.*

7.3.6 Approximation of the reference system behavior

This section shows that the deviation between the behavior of the reference system (7.23), (7.24) and the control system (7.1a), (7.1b) with the distributed event-based state feedback (7.28), (7.35) is bounded and the maximum deviation can be adjusted by means of an appropriate choice of the event threshold vectors $\bar{e}_j$ for all $j \in \mathcal{N}$. This result is used to determine the asymptotically stable set $\mathcal{A}_i$ for each subsystem Σ_i.

Recall the general condition for the boundedness of the deviation between the behavior of the event-based control system and the corresponding reference system, given in Theorem (4.5). Transferred to the context of distributed event-based control this condition claims that the difference states $x^i_{\Delta j}(t)$ must be bounded for all $j \in \mathcal{P}_i$. According to (7.29) this condition is fulfilled here due to the event triggering mechanism which induces transmissions and requests of information used to reset the states of the models (7.28), (7.30).

The following theorem makes this result more concrete and gives an upper bound for the deviation of the behavior of the event-based control system (7.1a), (7.1b), (7.28), (7.35) and the reference system (7.23), (7.24).

Theorem 7.3 *Consider the event-based control system (7.1a), (7.1b), (7.28), (7.35) and the reference system (7.23), (7.24) with the plant states $\boldsymbol{x}(t)$ and $\boldsymbol{x}_{\mathrm{r}}(t)$, respectively. The deviation $\mathbf{e}(t) = \boldsymbol{x}(t) - \boldsymbol{x}_{\mathrm{r}}(t)$ is bounded from above by*

$$|\mathbf{e}(t)| \leq \boldsymbol{e}_{\max} := \int_0^{\infty} \left| \mathrm{e}^{\bar{\boldsymbol{A}}t} \boldsymbol{B} \right| \mathrm{d}t \cdot |\bar{\boldsymbol{K}}| \begin{pmatrix} \bar{\boldsymbol{e}}_1 \\ \vdots \\ \bar{\boldsymbol{e}}_N \end{pmatrix} \tag{7.39}$$

for all $t \geq 0$ with the matrix $\bar{\boldsymbol{K}}$ given in (7.27).

Proof. The subsystem Σ_i represented by (7.1a) with the control input that is generated by the control input generator (7.28) is described by the state-space model

$$\begin{aligned}
\dot{\boldsymbol{x}}_i(t) &= \bar{\boldsymbol{A}}_i \boldsymbol{x}_i(t) - \boldsymbol{B}_i \sum_{j \in \mathcal{P}_i} \boldsymbol{K}_{ij} \tilde{\boldsymbol{x}}^i_j(t) + \boldsymbol{E}_i \boldsymbol{d}_i(t) + \boldsymbol{E}_{\mathrm{s}i} \boldsymbol{s}_i(t), \quad \boldsymbol{x}_i(0) = \boldsymbol{x}_{0i} \\
\boldsymbol{z}_i(t) &= \boldsymbol{C}_{\mathrm{z}i} \boldsymbol{x}_i(t)
\end{aligned}$$

with the matrix $\bar{\boldsymbol{A}}_i = (\boldsymbol{A}_j - \boldsymbol{B}_i \boldsymbol{K}_{\mathrm{d}i})$. The state $\boldsymbol{x}_i(t)$ depends upon the difference states $\boldsymbol{x}^i_{\Delta j}(t) = \boldsymbol{x}_j(t) - \tilde{\boldsymbol{x}}^i_j(t)$ for all $j \in \mathcal{P}_i$ which can be seen by reformulating the previous model as

$$\dot{\boldsymbol{x}}_i(t) = \bar{\boldsymbol{A}}_i\boldsymbol{x}_i(t) - \boldsymbol{B}_i \sum_{j\in\mathcal{P}_i} \boldsymbol{K}_{ij}\boldsymbol{x}_j(t) + \boldsymbol{B}_i \sum_{j\in\mathcal{P}_i} \boldsymbol{K}_{ij}\big(\boldsymbol{x}_j(t) - \tilde{\boldsymbol{x}}_j^i(t)\big) + \boldsymbol{E}_i\boldsymbol{d}_i(t) + \boldsymbol{E}_{\mathrm{s}i}\boldsymbol{s}_i(t) \tag{7.40a}$$

$$\boldsymbol{x}_i(0) = \boldsymbol{x}_{0i} \tag{7.40b}$$

$$\boldsymbol{z}_i(t) = \boldsymbol{C}_{\mathrm{z}i}\boldsymbol{x}_i(t). \tag{7.40c}$$

The interconnection of the models (7.40) according to the relation (7.1b) yields the overall control system model

$$\dot{\boldsymbol{x}}(t) = \bar{\boldsymbol{A}}\boldsymbol{x}(t) + \boldsymbol{E}\boldsymbol{d}(t) + \boldsymbol{B}\begin{pmatrix} \sum_{j\in\mathcal{P}_1} \boldsymbol{K}_{1j}\tilde{\boldsymbol{x}}_{\Delta j}^1(t) \\ \vdots \\ \sum_{j\in\mathcal{P}_N} \boldsymbol{K}_{Nj}\tilde{\boldsymbol{x}}_{\Delta j}^N(t) \end{pmatrix}, \quad \boldsymbol{x}(0) = \boldsymbol{x}_0, \tag{7.41}$$

where $\bar{\boldsymbol{A}} = \boldsymbol{A} - \boldsymbol{B}\left(\boldsymbol{K}_{\mathrm{d}} + \bar{\boldsymbol{K}}\right)$ with the state-feedback gains $\boldsymbol{K}_{\mathrm{d}}$ and $\bar{\boldsymbol{K}}$ as given in (7.26) and (7.27), respectively, and the matrices $\boldsymbol{A}$, $\boldsymbol{B}$, $\boldsymbol{E}$ according to (2.6).

Now consider the deviation $\boldsymbol{e}(t) = \boldsymbol{x}(t) - \boldsymbol{x}_{\mathrm{r}}(t)$. With Eqs. (7.25), (7.41) the deviation $\boldsymbol{e}(t)$ is described by the model

$$\dot{\boldsymbol{e}}(t) = \bar{\boldsymbol{A}}\boldsymbol{e}(t) + \boldsymbol{B}\begin{pmatrix} \sum_{j\in\mathcal{P}_1} \boldsymbol{K}_{1j}\tilde{\boldsymbol{x}}_{\Delta j}^1(t) \\ \vdots \\ \sum_{j\in\mathcal{P}_N} \boldsymbol{K}_{Nj}\tilde{\boldsymbol{x}}_{\Delta j}^N(t) \end{pmatrix}, \quad \boldsymbol{e}(0) = \boldsymbol{0}$$

which yields

$$\boldsymbol{e}(t) = \int_0^t \mathrm{e}^{\bar{\boldsymbol{A}}(t-\tau)}\boldsymbol{B}\begin{pmatrix} \sum_{j\in\mathcal{P}_1} \boldsymbol{K}_{1j}\tilde{\boldsymbol{x}}_{\Delta j}^1(\tau) \\ \vdots \\ \sum_{j\in\mathcal{P}_N} \boldsymbol{K}_{Nj}\tilde{\boldsymbol{x}}_{\Delta j}^N(\tau) \end{pmatrix} \mathrm{d}\tau.$$

The last equation is used to derive a bound on the maximum deviation $|\boldsymbol{e}(t)|$ by means of the following estimation

$$|\boldsymbol{e}(t)| \leq \int_0^t \left|\mathrm{e}^{\bar{\boldsymbol{A}}(t-\tau)}\boldsymbol{B}\right| \begin{pmatrix} \sum_{j\in\mathcal{P}_1} |\boldsymbol{K}_{1j}|\,|\tilde{\boldsymbol{x}}_{\Delta j}^1(\tau)| \\ \vdots \\ \sum_{j\in\mathcal{P}_N} |\boldsymbol{K}_{Nj}|\,|\tilde{\boldsymbol{x}}_{\Delta j}^N(\tau)| \end{pmatrix} \mathrm{d}\tau$$

$$\leq \int_0^\infty \left| \mathrm{e}^{\bar{\boldsymbol{A}}t} \boldsymbol{B} \right| \mathrm{d}t \begin{pmatrix} \sum_{j\in\mathcal{P}_1} |\boldsymbol{K}_{1j}| \, \bar{\boldsymbol{e}}_j \\ \vdots \\ \sum_{j\in\mathcal{P}_N} |\boldsymbol{K}_{Nj}| \, \bar{\boldsymbol{e}}_j \end{pmatrix} =: \boldsymbol{e}_{\max} \tag{7.42}$$

where the relation (7.34) (for each $j \in \mathcal{N}$) has been applied. Consequently, Eq. (7.42) can be reformulated as (7.39), which completes the proof. □

Theorem 7.3 shows that the system (7.1a), (7.1b) with the distributed event-based state feedback (7.28), (7.35) can be made to approximate the behavior of the reference system (7.23), (7.24) with adjustable accuracy, by appropriately setting the event thresholds $\bar{\boldsymbol{e}}_i$ for all $i \in \mathcal{N}$. Interestingly, the weighting factors α_i $(i \in \mathcal{N})$ that balance the event threshold vector $\bar{\boldsymbol{e}}_i$ in the transmit- and request-conditions (7.31), (7.32) do not affect the maximum deviation (7.39) in any way.

Note that the bound (7.39) holds element-wise and, thus, implicitly expresses also a bound on the deviation

$$\boldsymbol{e}_i(t) = \boldsymbol{x}_i(t) - \boldsymbol{x}_{\mathrm{r}i}(t)$$

between the subsystem states of the event-based control system and the corresponding reference subsystem. From (7.39) the bound

$$|\boldsymbol{e}_i(t)| \leq \boldsymbol{e}_{\max i} = \begin{pmatrix} \boldsymbol{O}_{n_i \times n_1} & \dots & \boldsymbol{O}_{n_i \times n_{i-1}} & \boldsymbol{I}_{n_i} & \boldsymbol{O}_{n_i \times n_{i+1}} & \dots & \boldsymbol{O}_{n_i \times n_N} \end{pmatrix} \boldsymbol{e}_{\max} \tag{7.43}$$

follows.

The result (7.43) can be combined with the result (7.20), (7.21) on the asymptotically stable set $\mathcal{A}_{\mathrm{r}i}$ for the reference systems (7.23) in order to infer the asymptotically stable set $\mathcal{A}_i$ for the subsystems Σ_i with distributed event-based state feedback.

Corollary 7.1 *Consider the event-based control system* (7.1a), (7.1b), (7.28), (7.35). *Each subsystem Σ_i is practically stable with respect to the set*

$$\mathcal{A}_i := \left\{ \boldsymbol{x}_i \in \mathbb{R}^{n_i} \;\middle|\; |\boldsymbol{x}_i| \leq \boldsymbol{b}_i \right\} \tag{7.44}$$

with the bound

$$\boldsymbol{b}_i = \boldsymbol{b}_{\mathrm{r}i} + \boldsymbol{e}_{\max i} \tag{7.45}$$

where $\boldsymbol{b}_{\mathrm{r}i}$, given in (7.21), *is the bound for the reference subsystem* (7.23) *and $\boldsymbol{e}_{\max i}$, given in* (7.43), *denotes the maximum deviation between the behavior of subsystem Σ_i subject to the distributed event-based controller* (7.28), (7.35) *and the reference subsystem* (7.23).

The Corollary 7.1 can be interpreted as follows: Given that the vector $\boldsymbol{e}_{\max}$ can be freely manipulated by the choice of the event threshold vectors $\bar{\boldsymbol{e}}_i$ for all $i \in \mathcal{N}$, the difference between the sets $\mathcal{A}_{\mathrm{r}i}$ and $\mathcal{A}_i$ can be arbitrarily adjusted, as well.

7.3.7 Minimum inter-event times

This section extends the results given in Propositions 7.1, 7.2 and determines bounds on the minimum inter-event times for events that are triggered by the transmit-condition (7.31) and the request-condition (7.32). The following analysis is based on the assumption that the state $\boldsymbol{x}_i(t)$ of subsystem Σ_i is contained within the set $\mathcal{A}_i$ for all $i \in \mathcal{N}$.

The following theorem states that the minimum inter-event time

$$T_{\mathrm{min,TC}i} := \min_{k_i} \left(t_{k_i+1} - t_{k_i}\right), \quad \forall\, k_i \in \mathbb{N}_0 \tag{7.46}$$

for two consecutive events that are triggered according to the transmit-condition (7.31) by the controller F_i is bounded from below by some time $\bar{T}_{\mathrm{TC}i}$.

Theorem 7.4 *The minimum inter-event time $T_{\mathrm{min,TC}i}$ defined in (7.46) is bounded from below by*

$$T_{\mathrm{min,TC}i} \geq \bar{T}_{\mathrm{TC}i}, \quad \forall\, t \geq 0,$$

where the time $\bar{T}_{\mathrm{TC}i}$ is given by

$$\bar{T}_{\mathrm{TC}i} := \arg\min_t \left\{ \int_0^t \left| \mathrm{e}^{\bar{\boldsymbol{A}}_i \tau} \right| \mathrm{d}\tau \left(\sum_{j \in \mathcal{P}_i} |\boldsymbol{B}_i \boldsymbol{K}_{ij}| \, (\boldsymbol{b}_j + \bar{\boldsymbol{e}}_j) + \sum_{j \in \mathcal{P}_i} |\boldsymbol{E}_{\mathrm{s}i} \boldsymbol{L}_{ij} \boldsymbol{C}_{\mathrm{z}j}| \, \boldsymbol{b}_j + |\boldsymbol{E}_i| \, \bar{\boldsymbol{d}}_i \right) = \alpha_i \bar{\boldsymbol{e}}_i \right\} \tag{7.47}$$

with the bounds $\boldsymbol{b}_j$ given in (7.45).

Proof. Consider Eq. (7.36) with $\hat{t}_{\mathrm{TX}i} = t_{k_i}$, which yields

$$\boldsymbol{\delta}_{\mathrm{c}i}(t) = \int_{t_{k_i}}^t \mathrm{e}^{\bar{\boldsymbol{A}}_i (t-\tau)} \left(- \sum_{j \in \mathcal{P}_i} \boldsymbol{B}_i \boldsymbol{K}_{ij} \tilde{\boldsymbol{x}}_j^i(\tau) + \boldsymbol{E}_i \boldsymbol{d}_i(\tau) + \boldsymbol{E}_{\mathrm{s}i} \boldsymbol{s}_i(\tau) \right) \mathrm{d}\tau.$$

With Eq. (7.1b) and the relation $z_j(t) = C_{zj}x_j(t)$, the previous equation can be restated as

$$\delta_{ci}(t) = \int_{t_{k_i}}^{t} e^{\bar{A}_i(t-\tau)} \left(-\sum_{j\in\mathcal{P}_i} B_i K_{ij} \tilde{x}_j^i(\tau) + E_i d_i(\tau) + \sum_{j\in\mathcal{P}_i} E_{si} L_{ij} C_{zj} x_j(\tau) \right) d\tau.$$

In order to analyze the minimum time that elapses in between the two consecutive events k_i and k_i+1, assume that no information is requested from F_i after t_{k_i} until the next event at time t_{k_i+1} occurs. Recall that an event is triggered according to the transmit-condition (7.31) whenever $|\delta_{ci}(t)| = |x_i(t) - x_{ci}(t)| = \alpha_i \bar{e}_i$ holds. Hence, the minimum inter-event time $T_{\min,TCi}$ is the minimum time t for which

$$\left| \int_{0}^{t} e^{\bar{A}_i(t-\tau)} \left(-\sum_{j\in\mathcal{P}_i} B_i K_{ij} \tilde{x}_j^i(\tau) + E_i d_i(\tau) + \sum_{j\in\mathcal{P}_i} E_{si} L_{ij} C_{zj} x_j(\tau) \right) d\tau \right| = \alpha_i \bar{e} \quad (7.48)$$

is satisfied. The following analysis derives a bound on that minimum inter-event time. To this end, it is assumed that the state $x_i(t)$ is bounded to the set $\mathcal{A}_i$ given in (7.44) for all $i \in \mathcal{N}$ which implies that

$$|x_i(t)| \leq b_i \quad (7.49)$$

holds where the bound b_i is given in (7.45). By virtue of the event triggering mechanism and the state reset the prediction $\tilde{x}_j^i(t)$ always remains in a bounded surrounding of the subsystems state $x_j(t)$. Hence, from Eq. (7.34) the relation

$$\left|\tilde{x}_j^i(t)\right| \leq |x_j(t)| + \bar{e}_j \leq b_j + \bar{e}_j \quad (7.50)$$

follows. Taking account of the bound (2.5) on the disturbance $d_i(t)$, the left-hand side of Eq. (7.48) is bounded by

$$\begin{aligned} &\left| \int_{0}^{t} e^{\bar{A}_i(t-\tau)} \left(-\sum_{j\in\mathcal{P}_i} B_i K_{ij} \tilde{x}_j^i(\tau) + E_i d_i(\tau) + \sum_{j\in\mathcal{P}_i} E_{si} L_{ij} C_{zj} x_j(\tau) \right) d\tau \right| \\ &\quad \leq \left| \int_{0}^{t} e^{\bar{A}_i \tau} \right| d\tau \left(\sum_{j\in\mathcal{P}_i} |B_i K_{ij}| (b_j + \bar{e}_j) + \sum_{j\in\mathcal{P}_i} |E_{si} L_{ij} C_{zj}| \, b_j + |E_i| \, \bar{d}_i \right). \end{aligned}$$

Consequently, for the minimum inter-event time $T_{\min,TCi}$ the relation

$$\bar{T}_{TCi} \leq T_{\min,TCi}$$

holds where

$$\bar{T}_{\mathrm{TC}i} := \arg\min_t \left\{ \int_0^t \left| \mathrm{e}^{\bar{\boldsymbol{A}}_i \tau} \right| \mathrm{d}\tau \left(\sum_{j \in \mathcal{P}_i} |\boldsymbol{B}_i \boldsymbol{K}_{ij}| \, (\boldsymbol{b}_j + \bar{\boldsymbol{e}}_j) + \sum_{j \in \mathcal{P}_i} |\boldsymbol{E}_{\mathrm{s}i} \boldsymbol{L}_{ij} \boldsymbol{C}_{\mathrm{z}j}| \, \boldsymbol{b}_j + |\boldsymbol{E}_i| \, \bar{\boldsymbol{d}}_i \right) = \alpha_i \bar{\boldsymbol{e}}_i \right\},$$

which completes the proof. □

The next theorem presents a bound on the minimum inter-event time

$$T^i_{\min,\mathrm{RC}j} := \min_{r_i(j)} \left(t_{r_i(j)+1} - t_{r_i(j)} \right)$$

for two consecutive information requests from controller F_j, triggered according to the request-condition (7.32) by the controller F_i.

Theorem 7.5 *The minimum time $T^i_{\min,\mathrm{RC}j}$ that elapses in between two consecutive events at which the controller F_i requests information from controller F_j is bounded from below by*

$$T^i_{\min,\mathrm{RC}j} \geq \bar{T}^i_{\mathrm{RC}j}, \quad \forall\, t \geq 0,$$

the time $\bar{T}^i_{\mathrm{RC}j}$ is given by

$$\bar{T}^i_{\mathrm{RC}j} := \arg\min_t \left\{ \int_0^t \left| \mathrm{e}^{\bar{\boldsymbol{A}}_j \tau} \right| \mathrm{d}\tau \left(\sum_{p \in \mathcal{P}_i \setminus \{j\}} |\boldsymbol{C}_{ji} \boldsymbol{A}_{\mathrm{a}i} \boldsymbol{T}_{ip}| \, (\boldsymbol{b}_p + \bar{\boldsymbol{e}}_p) + |\boldsymbol{C}_{ji} \boldsymbol{B}_{\mathrm{a}i} \boldsymbol{C}_{\mathrm{z}i}| \, \boldsymbol{b}_i \right) = (1 - \alpha_j) \bar{\boldsymbol{e}}_j \right\} \quad (7.51)$$

with the bounds $\boldsymbol{b}_i$ as in (7.45).

Proof. Consider Eq. (7.38) with $\hat{t}_{\mathrm{Rx}i(j)} = t_{r_i(j)}$, which results in

$$\boldsymbol{\delta}^i_{\mathrm{c}j}(t) = \int_{t_{r_i(j)}}^t \mathrm{e}^{\bar{\boldsymbol{A}}_j (t-\tau)} \left(\sum_{p \in \mathcal{P}_i \setminus \{j\}} \boldsymbol{C}_{ji} \boldsymbol{A}_{\mathrm{a}i} \boldsymbol{T}_{ip} \tilde{\boldsymbol{x}}^i_p(\tau) + \boldsymbol{C}_{ji} \boldsymbol{B}_{\mathrm{a}i} \boldsymbol{C}_{\mathrm{z}i} \boldsymbol{x}_i(\tau) \right) \mathrm{d}\tau.$$

In the following, it is assumed that the controller F_j does not trigger an event according to its transmission-condition (7.31) in between the times $t_{r_i(j)}$ and $t_{r_i+1(j)}$ in order to identify a minimum time that elapses in between two consecutive events that trigger the request of information from F_j by the controller F_i. Note that this information request is triggered whenever the equality $\left|\boldsymbol{\delta}^i_{cj}(t)\right| = |\boldsymbol{C}_{ji}\tilde{\boldsymbol{x}}_{ai}(t) - \boldsymbol{x}_{cj}(t)| = (1-\alpha_j)\bar{\boldsymbol{e}}_j$ is satisfied in at least one element. The minimum inter-event time $T^i_{\min,\mathrm{RC}j}$ is given by the minimum time t for which

$$\left|\int_0^t \mathrm{e}^{\bar{\boldsymbol{A}}_j(t-\tau)}\left(\sum_{p\in\mathcal{P}_i\setminus\{j\}} \boldsymbol{C}_{ji}\boldsymbol{A}_{ai}\boldsymbol{T}_{ip}\tilde{\boldsymbol{x}}^i_p(\tau) + \boldsymbol{C}_{ji}\boldsymbol{B}_{ai}\boldsymbol{C}_{zi}\boldsymbol{x}_i(\tau)\right)\mathrm{d}\tau\right| = 1-\alpha_j\bar{\boldsymbol{e}}_j$$

holds. Following the same arguments as in the previous proof, a bound

$$\bar{T}^i_{\mathrm{RC}j} \leq T^i_{\min,\mathrm{RC}j}$$

on that minimum inter-event time is given by

$$\bar{T}^i_{\mathrm{RC}j} := \arg\min_t\left\{\int_0^t \left|\mathrm{e}^{\bar{\boldsymbol{A}}_j\tau}\right|\mathrm{d}\tau\left(\sum_{p\in\mathcal{P}_i\setminus\{j\}} |\boldsymbol{C}_{ji}\boldsymbol{A}_{ai}\boldsymbol{T}_{ip}|\,(\boldsymbol{b}_p+\bar{\boldsymbol{e}}_p)\right.\right.$$
$$\left.\left. + |\boldsymbol{C}_{ji}\boldsymbol{B}_{ai}\boldsymbol{C}_{zi}|\,\boldsymbol{b}_i\right) = (1-\alpha_j)\bar{\boldsymbol{e}}_j\right\},$$

which concludes the proof. □

7.4 Example: Distributed event-based control of the thermofluid process

This section illustrates the behavior of the distributed event-based state-feedback approach for the example of the thermofluid process introduced in Sec. 2.4. Subsequently, the process is considered to consist of four interconnected subsystems which are described in Appendix A.2. Moreover, it is demonstrated how the distributed state feedback is designed according to the Algorithm 7.1 and how the obtained control law is implemented in an event-based fashion. The results of a simulation and an experiment are shown at the end of this section.

7.4.1 Design of the distributed event-based controller

This section shows how a state-feedback law is determined first which is afterwards implemented in an event-based manner.

Distributed state-feedback design. According to step 1 in the Algorithm 7.1, an approximate model $\Sigma_{\mathrm{a}i}$ together with the bounds $\bar{\boldsymbol{G}}_{\mathrm{fv}i}(t)$ and $\bar{\boldsymbol{G}}_{\mathrm{fd}i}(t)$ on the impulse response matrices for the residual models $\Sigma_{\mathrm{f}i}$ must be determined for each subsystem Σ_i $(i = 1, \dots, 4)$. For the example of the thermofluid process, the approximate models with the corresponding residual models are presented in Appendix A.3. Given the approximate models $\Sigma_{\mathrm{a}i}$ and the subsystems Σ_i, the parameters of the extended subsystem models $\Sigma_{\mathrm{e}i}$ result from Eq. (7.10).

In the second step, the state-feedback gains $\boldsymbol{K}_{\mathrm{e}i}$ are to be designed such that the controlled extended subsystems have a desired behavior. In this example these matrices are chosen as follows:

$$\begin{aligned} \boldsymbol{K}_{\mathrm{e}1} &= \begin{pmatrix} 3.00 & 1.05 \end{pmatrix}, \quad \boldsymbol{K}_{\mathrm{e}2} = \begin{pmatrix} -0.70 & 0.89 & -1.13 & -0.14 \end{pmatrix}, \\ \boldsymbol{K}_{\mathrm{e}3} &= \begin{pmatrix} 1.00 & 1.10 \end{pmatrix}, \quad \boldsymbol{K}_{\mathrm{e}4} = \begin{pmatrix} 0.60 & -1.33 & 0.16 & 1.12 \end{pmatrix}. \end{aligned} \tag{7.52}$$

The structure of the distributed state-feedback law is better conveyed by restating these matrices in the form used in Eq. (7.16) which yields

$$\boldsymbol{K}_{\mathrm{d}1} = 3.00, \qquad \boldsymbol{K}_{13} = 1.05, \tag{7.53a}$$

$$\boldsymbol{K}_{21} = 0.89, \quad \boldsymbol{K}_{\mathrm{d}2} = -0.70, \quad \boldsymbol{K}_{23} = -1.13, \quad \boldsymbol{K}_{24} = -0.14, \tag{7.53b}$$

$$\boldsymbol{K}_{31} = 1.10, \qquad \boldsymbol{K}_{\mathrm{d}3} = 1.00, \tag{7.53c}$$

$$\boldsymbol{K}_{41} = -1.33, \quad \boldsymbol{K}_{42} = 0.16, \quad \boldsymbol{K}_{43} = 1.12, \quad \boldsymbol{K}_{\mathrm{d}4} = 0.60. \tag{7.53d}$$

Note that the structure of the overall state-feedback gain is implicitly predetermined in the proposed design method by the interconnections of the subsystems. For the considered system, this design procedure leads to the fact that the controllers F_2 and F_4 of subsystems Σ_2 or Σ_4, respectively, use state information of all remaining subsystems.

In the final step, the stability of the overall control system is analyzed by means of the condition (7.19). The analysis results are summarized as follows:

$$\lambda_{\mathrm{P}}\left(\int_0^\infty \bar{\boldsymbol{G}}_{\mathrm{fv1}}(t)\mathrm{d}t \int_0^\infty \left|\boldsymbol{H}_{\mathrm{e}i}\mathrm{e}^{\bar{\boldsymbol{A}}_{\mathrm{e1}}t}\boldsymbol{F}_{\mathrm{e1}}\right|\mathrm{d}t\right) = 1.48 \cdot 10^{-4} < 1,$$
$$\lambda_{\mathrm{P}}\left(\int_0^\infty \bar{\boldsymbol{G}}_{\mathrm{fv2}}(t)\mathrm{d}t \int_0^\infty \left|\boldsymbol{H}_{\mathrm{e2}}\mathrm{e}^{\bar{\boldsymbol{A}}_{\mathrm{e2}}t}\boldsymbol{F}_{\mathrm{e2}}\right|\mathrm{d}t\right) = 0.37 < 1,$$
$$\lambda_{\mathrm{P}}\left(\int_0^\infty \bar{\boldsymbol{G}}_{\mathrm{fv3}}(t)\mathrm{d}t \int_0^\infty \left|\boldsymbol{H}_{\mathrm{e3}}\mathrm{e}^{\bar{\boldsymbol{A}}_{\mathrm{e3}}t}\boldsymbol{F}_{\mathrm{e3}}\right|\mathrm{d}t\right) = 2.51 \cdot 10^{-5} < 1,$$
$$\lambda_{\mathrm{P}}\left(\int_0^\infty \bar{\boldsymbol{G}}_{\mathrm{fv4}}(t)\mathrm{d}t \int_0^\infty \left|\boldsymbol{H}_{\mathrm{e4}}\mathrm{e}^{\bar{\boldsymbol{A}}_{\mathrm{e4}}t}\boldsymbol{F}_{\mathrm{e4}}\right|\mathrm{d}t\right) = 0.37 < 1.$$

Consequently, the distributed state-feedback controller (7.53) guarantees the stability of the overall control system. According to Theorem 7.2, each subsystem is practically stable with respect the set $\mathcal{A}_{\mathrm{r}i}$ defined in (7.20) with the bounds

$$b_{\mathrm{r1}} = 0.00, \quad b_{\mathrm{r2}} = 1.20, \quad b_{\mathrm{r3}} = 6.11, \quad b_{\mathrm{r4}} = 0.31. \tag{7.54}$$

The result $b_{\mathrm{r1}} = 0$ shows that in the distributed continuous state-feedback system the disturbance and couplings do not affect state $x_1(t)$ and, thus, this state converges to the origin for $t \to \infty$.

Event-based implementation of the distributed state feedback. For the implementation of the derived state-feedback in an event-based manner, the event thresholds $\bar{e}_i$ and the weighting factor α_i for $i = 1, \ldots, 4$ are chosen as follows:

$$\begin{aligned} \bar{e}_1 = \bar{e}_3 = 0.035, &\qquad \alpha_1 = \alpha_3 = 0.85, \\ \bar{e}_2 = \bar{e}_4 = 3.2, &\qquad \alpha_2 = \alpha_4 = 0.94. \end{aligned} \tag{7.55}$$

According to Theorem 7.3, the maximum deviation between the distributed event-based state-feedback systems and the continuous reference system results in

$$\boldsymbol{e}_{\max} = \begin{pmatrix} 0.67 & 0.51 & 1.31 & 0.79 \end{pmatrix}^{\mathrm{T}}. \tag{7.56}$$

By virtue of Corollary 7.1, this result together with the bounds (7.54) implies the practical stability of the subsystems Σ_i with respect to the sets $\mathcal{A}_i$ given in (7.44) with

$$b_1 = 0.67, \quad b_2 = 1.71, \quad b_3 = 7.43, \quad b_4 = 1.10. \tag{7.57}$$

The following table summarizes the results on the bounds $\bar{T}_{\mathrm{TC}i}$ and $\bar{T}^i_{\mathrm{RC}j}$ on the minimum inter-event times for events triggered due to the transmit-condition (7.31) or the request-condition (7.32). These times are obtained according to Theorem 7.4 and Theorem 7.5. The '–' symbolizes that Σ_j is not a predecessor of Σ_i and, hence, F_i never requests information from F_j.

Table 7.1: Minimum inter-event times (MIET)

F_i	MIET $\bar{T}_{\mathrm{TC}i}$ in s for transmissions	MIET $\bar{T}^i_{\mathrm{RC}j}$ in s for requests of information from F_j			
		$j=1$	$j=2$	$j=3$	$j=4$
F_1	0.83	–	–	2.78	–
F_2	4.21	0.29	–	2.64	0.58
F_3	6.97	0.29	–	–	–
F_4	4.62	0.29	5.97	0.65	–

The presented analysis results are evaluated by means of a simulation and an example in the next section.

7.4.2 Simulation and experimental results

The following presents the results of a simulation and an experiment, demonstrating the disturbance behavior of the distributed event-based state-feedback approach.

Simulation results. The simulation investigates the scenario where the overall system is perturbed in the time intervals $t \in [200, 600]$ s by the disturbance $d_{\mathrm{H}}(t)$ and in $t \in [1000, 1400]$ s by the disturbance $d_{\mathrm{F}}(t)$. The behavior of the overall plant with the distributed event-based controller subject to these disturbances is illustrated in Fig. 7.7. It shows the disturbance behavior of each subsystem where from top to bottom the following signals are depicted: the disturbance $d_{\mathrm{H}}(t)$ or $d_{\mathrm{F}}(t)$, the subsystem state $x_i(t)$, the control input $u_i(t)$ and the time instants where information is received (Rx_i) and transmitted (Tx_i), indicated by stems. Regarding the reception of information the amplitude of the stems refers to the subsystem that has transmitted the information. For both Rx_i and Tx_i a stem with a circle denotes that the request-condition (7.32) or the transmit-condition (7.31) of the controller F_i is satisfied and, thus, leads to the triggering of the respective event. For the receptions (Rx_i), a stem with a diamond indicates that another

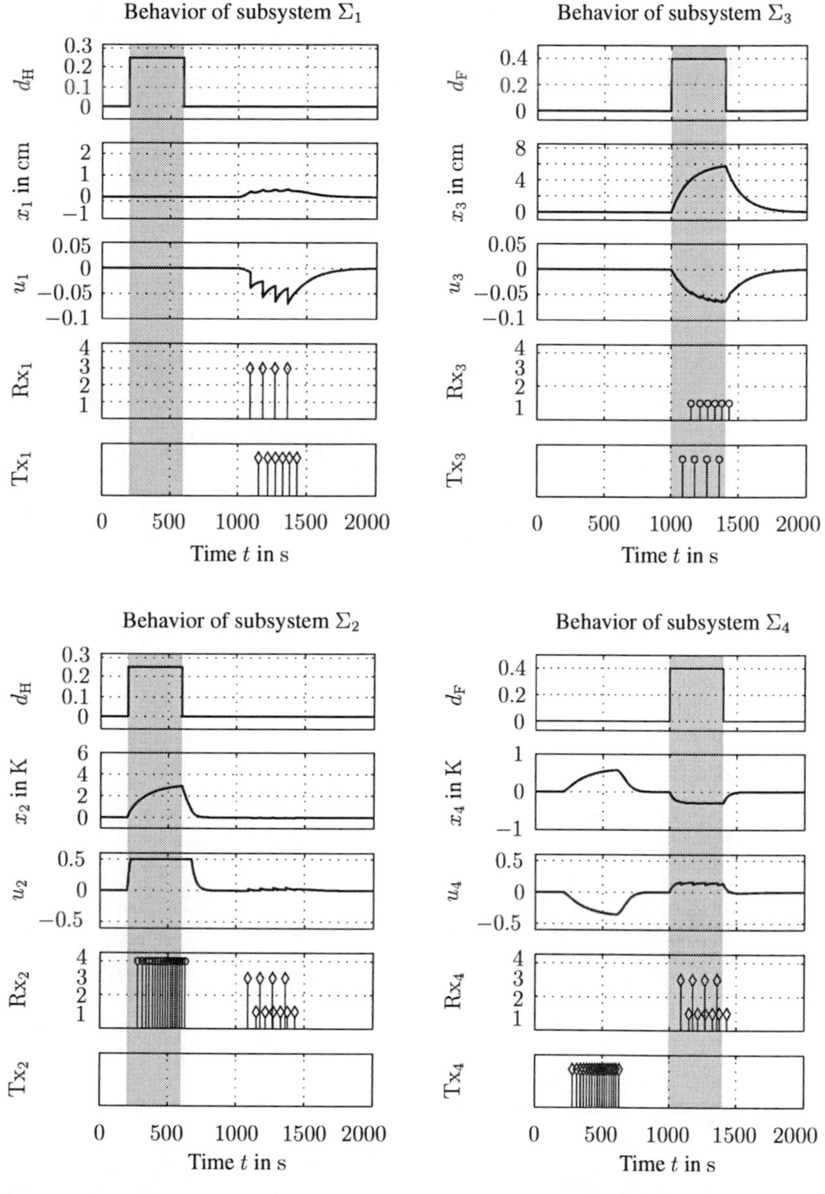

Figure 7.7: Simulation results for the disturbance behavior of the distributed event-based state feedback.

control unit has transmitted information, because its local transmit-condition was satisfied or its state was requested by a third controller. For the transmissions (Tx_i), a stem with a diamond shows that the controller F_i was requested to transmit the local state by another controller.

The simulation results show that the communication is adapted to the behavior of the system. No event is triggered when the overall system is undisturbed, except for the times immediately after the disturbances vanish. In this simulation the transmit-condition (7.31) is only satisfied in the controller F_3, whereas the request of information is only induced by the controllers F_2 and F_3. In this investigation the minimum inter-event times for events that are triggered according to the transmit-condition (7.31) or the request-condition (7.32) are

$$\begin{aligned} T_{\min,\mathrm{TC}3} &\approx 90\,\mathrm{s} \;>\; 6.97\,\mathrm{s} = \bar{T}_{\mathrm{TC}3}, \\ T^2_{\min,\mathrm{RC}4} &\approx 0.7\,\mathrm{s} \;>\; 0.58\,\mathrm{s} = \bar{T}^2_{\mathrm{RC}4}, \\ T^3_{\min,\mathrm{RC}1} &\approx 49\,\mathrm{s} \;>\; 0.29\,\mathrm{s} = \bar{T}^3_{\mathrm{RC}1}. \end{aligned}$$

For these inter-event times the bounds that are listed in Tab. 7.1 hold true. Note that in the time interval $t \in [200, 600]\,\mathrm{s}$ a large number of events is triggered by the controller F_2 that keeps requesting information from the controller F_4. The high rate of the event triggering is due to the fact that in this interval the control input $u_2(t)$ is in saturation. The saturation of the control input is a non-linearity that is not considered in the linear model used by the controller F_2. In this way, this simulation shows how model uncertainties or unmodeled dynamics can lead to an increased triggering of events. But yet, compared to a conventional discrete-time controller with a sampling time of $h = 10\,\mathrm{s}$, the communication effort is considerable reduced

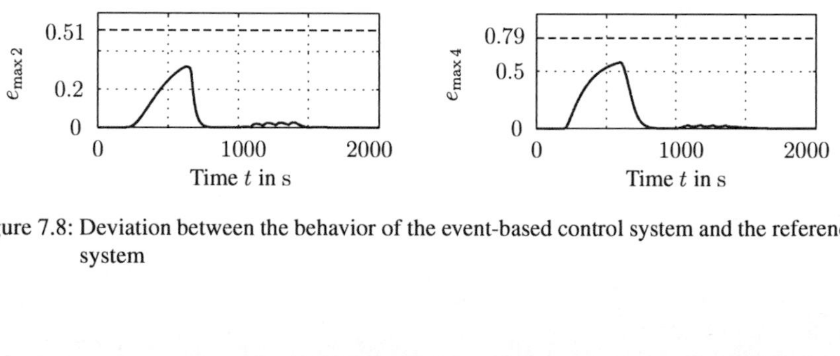

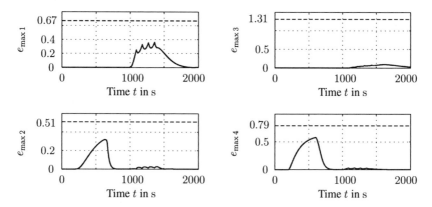

Figure 7.8: Deviation between the behavior of the event-based control system and the reference system

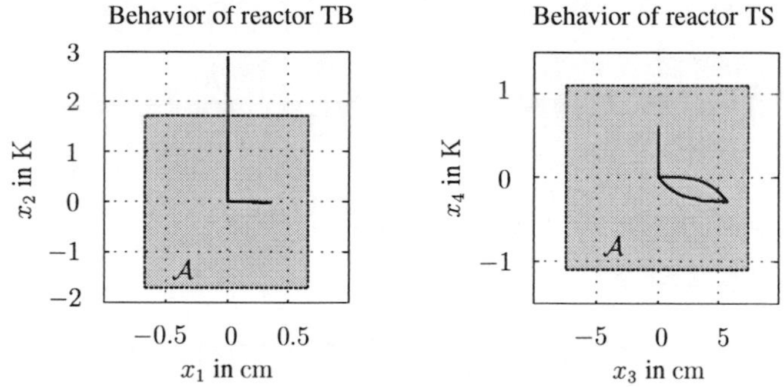

Figure 7.9: Simulated trajectory of the overall system state in state-space.

by the distributed event-based controller where only 28 information transmission are induced (either by the transmission-condition or by the request-condition of the control units) within the duration of 2000 s.

The deviation between the behavior of the event-based control system (7.1a), (7.1b), (7.28), (7.35) and the reference system (7.23), (7.24) is presented in Fig. 7.8. The dashed lines refer to maximum deviations $e_{\max i}$ as defined in (7.43) with $e_{\max}$ given in (7.56). This investigation verifies the analysis result (7.56) on the maximum deviation between the event-based and the continuous state-feedback system.

Figure 7.9 illustrates the trajectories of the subsystem states $x_i(t)$ in the state space. The dashed lines represent the bounds (7.57) which shape the set $\mathcal{A}$ to which the overall system state $\boldsymbol{x}(t)$ should be bounded according to the analysis which was done in the previous section. Note that the analytically determined bound $b_2 = 1.71$ does not hold for the state $x_2(t)$ which has the maximum amplitude

$$\max_{t \geq 0} |x_2(t)| = 2.9.$$

The discrepancy between the analysis and the simulation result can be explained by the saturation of the control input $u_2(t)$ around the time where the disturbance $d_{\mathrm{H}}(t)$ is active. With the exception of this phenomenon, the analysis result on the sets $\mathcal{A}_i$ hold true.

Experimental results. The following describes the experimental results obtained by the application of the distributed event-based state-feedback controller to the thermofluid process. The scenario considered for the experiment differs slightly from the one investigated in the simulation in that the disturbances are marginally smaller. The disturbance behavior is presented in Fig. 7.10 where the same signals are shown as in Fig. 7.7.

The experimental results are comparable with the outcome of the simulation although con-

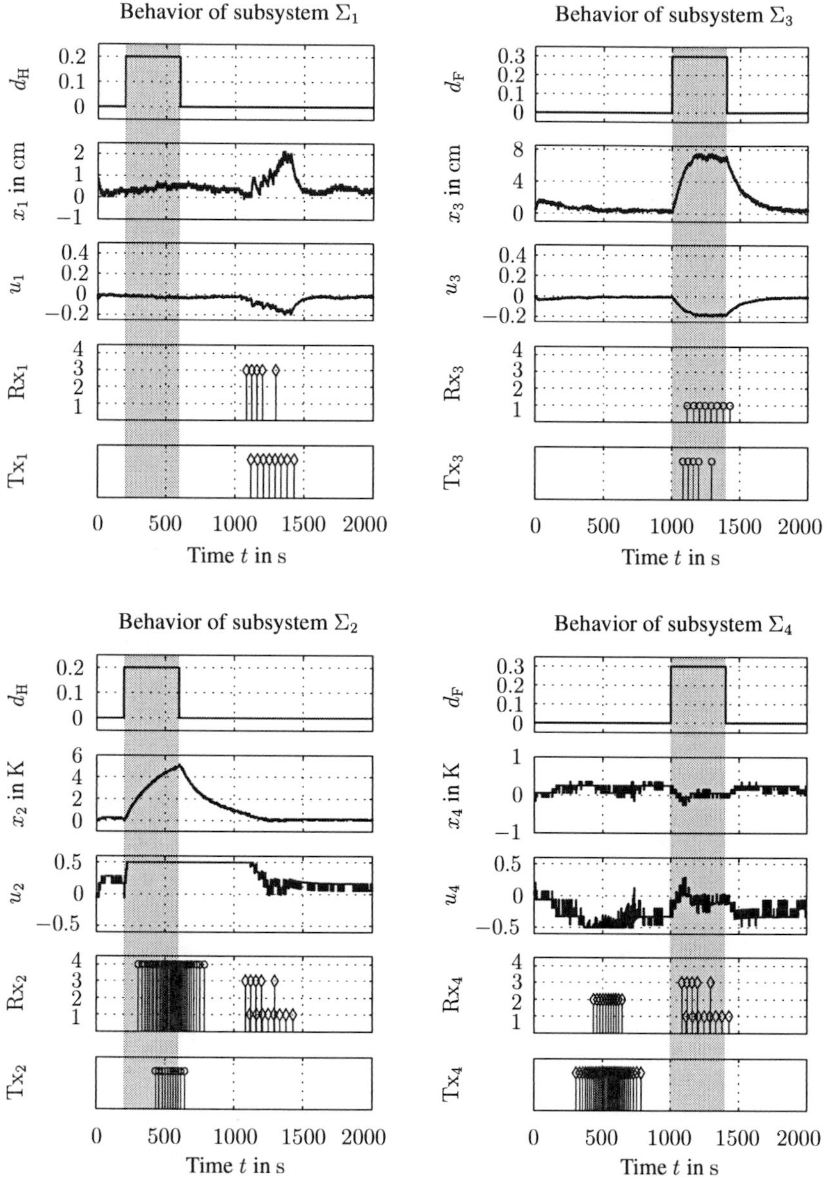

Figure 7.10: Experimental results for the disturbance behavior of the distributed event-based state feedback.

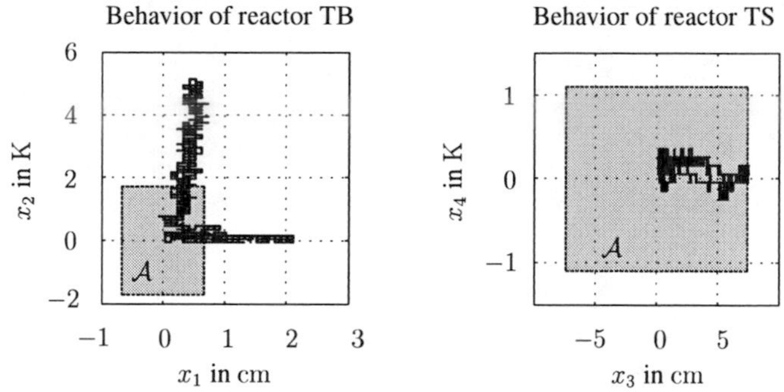

Figure 7.11: Trajectory of the overall system state in state-space in the experiment.

siderable more events are triggered. Nevertheless, the events are largely triggered in response to the disturbances that affect the overall system. Similarly to the simulation, the effect of the saturation of the control input $u_2(t)$ also occurs in the experiment and again results in the triggering of numerous events. In total, 60 events are triggered which cause a transmission of information within 2000 s. Hence, from the communication aspect, the distributed event-based controller outperforms a conventional discrete-time controller (with appropriately chosen sampling time $h = 10\,\mathrm{s}$). Moreover, the similarities between the experimental and the simulation results show that the proposed control approach is even robust with respect to model uncertainties.

Figure 7.11 evaluates to what extend the analysis results, presented in the previous section, are valid for trajectories that are measured in the experiment. It shows that the bounds $b_1 = 0.67$ and $b_2 = 1.71$ do not hold, whereas the bounds $b_3 = 7.43$ and $b_4 = 1.10$ are not exceeded by the states $x_3(t)$ or $x_4(t)$, respectively. The reason why $x_2(t)$ grows much larger than the bound b_2, is again the saturation of the control input $u_2(t)$. The exceeding of the state $x_1(t)$ above the bound b_1 can be ascribed to unmodeled nonlinear dynamics of various components of the plant.

In summary, the simulation and the experiment have shown that events are only triggered by the distributed event-based state-feedback controller whenever the plant is substantially affected by an exogenous disturbance and no feedback communication is required in the unperturbed case. This observation implies that the effect of the couplings between the subsystems has a minor impact on the triggering of events, since these dynamics are taken into account by using approximate models in the design of the local control units. In consequence, the proposed distributed event-based state-feedback method is suitable for the application to interconnected systems where neighboring subsystems are strongly coupled.

8 Summary and outlook

8.1 Contributions of the thesis

This thesis has presented three different methods for the event-based state-feedback control of physically interconnected systems. The common basis for all methods is a general structure of the event-based controller, consisting of local control units which exchange information over a communication network. The proposed event-based control approaches differ with respect to the applied model information in the control units and the communication effort. The main characteristic that all three methods share is that the disturbance behavior of a continuous state-feedback system is approximated by the respective event-based state-feedback system with arbitrary accuracy (Theorems 5.1, 6.1, 7.3) , while adapting the feedback communication to the system behavior.

In all three approaches, the control unit for each subsystem consists of a control input generator and an event generator. Both components include a dynamic model in order to determine a prediction of the respective subsystem state which is used for generating the control input or the events, respectively. The triggering of events causes the reset of the state prediction to the current subsystem state whenever the deviation between both reaches a predefined event threshold. In this way, the triggering mechanism yields the boundedness of the difference between the subsystem states and their predictions applied for the input generation which has been identified as the crucial property for accomplishing the approximation of the continuous reference systems behavior by the event-based control system (Section 4.1, Theorem 4.1).

The three proposed event-based control methods are characterized as follows:

1. In the distributed realization of the event-based state-feedback approach (Chapter 5) the model of the overall continuous state-feedback system is included in each component of the event-based controller and at the event times, information is broadcasted through the network. The method is suitable for implementing a state-feedback control with an arbitrarily structured feedback-gain in an event-based manner.

2. In the decentralized event-based state-feedback approach (Chapter 6) each control unit includes only local model information and communications are induced solely between

the event generator and the corresponding control input generator. In order to apply this method the overall system is required to be stabilizable by a decentralized state feedback.

3. In the distributed event-based state-feedback approach (Chapter 7) model information about the neighboring subsystems are incorporated in each control unit. The boundedness of the deviation between the subsystem states and their predictions is guaranteed by an event generator that triggers two types of events which induce the transmission of information to or the request of information from the neighboring subsystems.

Except for the first approach where complete model information is applied, the event-based control units do not only react to disturbances but also to the couplings that affect the subsystems, which has been explicitly taken into account by developing two methods for the coupling input estimation (Section 6.2.5) or by using model information about neighboring subsystems for the input generation (Sections 7.2.1, 7.3.4).

Physically interconnected systems are generally of large scale, which brings along the demand for controller design methods that require only partial model information. This thesis has presented two methods which use only local model information (Section 6.3) or information about neighboring subsystems (Section 7.3). The former requires the subsystems to be weakly interconnected (Theorem 6.3), while the latter allows the couplings among neighboring subsystems to be strong (Theorem 7.1).

When considering information transmissions over a communication network, the load of the network needs to be regarded. For all event-based control methods, the time that elapses in between the triggering of two consecutive events is bounded from below by a positive constant (Theorems 5.2, 6.6, 7.4, 7.5). The proposed control methods have been tested and evaluated on a thermofluid process in simulations and experiments (Sections 5.3, 6.2.6, 6.4, 7.4). These investigations haven shown that the novel event-based control approaches do not only guarantee minimum inter-event times but even significantly reduce the communication effort compared to a conventional discrete-time control. Moreover, the experiments have indicated that the proposed control approaches are robust with respect to model uncertainties.

8.2 Open problems

A further development of the theory on event-based control of interconnected systems should mainly focus on the following three aspects:

- The extension of the considered class of systems from linear to nonlinear systems, which better reflect the real behavior of technical systems in a larger range of validity than linear systems.

- The consideration of imperfect communication over real-time networks.
- The development of event-based control concepts that do not only aim at reducing the overall feedback communication over a large time interval, but which also avoid peaks in the network load due to the accumulation of a large number of events in short intervals.

These issues are discussed in the following in more detail.

Extensions to nonlinear systems. The extension of event-based control concepts to nonlinear physically interconnected systems is mainly motivated by its practical relevance. For single-loop systems an approach to input-output linearizable systems has been made by the author in [2, 3, 7] which can be developed further to cope with physically interconnected input-output linearizable systems. In literature there exist several works which deal with nonlinear event-based control which almost exclusively consider the most general class of nonlinear systems, [23, 47, 120, 149, 163, 165]. From a theoretical point of view the investigation of these systems is an interesting and challenging task which, however, often leads to very complex or even restrictive analysis methods, that are less relevant from a practical point of view, if the assumptions are hard to satisfy. In order to reduce the conservatism of existing results and to improve the applicability of the event-based control methods, further investigations should concentrate on systems that feature special aspects of nonlinearities. Such classes of nonlinear systems could be Wiener and Hammerstein systems, which are characterized by static nonlinearities in the output or input, respectively, or piecewise linear systems, [90], which have a larger range of validity than linear systems.

Imperfect communication. The main application field for event-based control are networked control systems where the feedback is closed over an unreliable real-time network. In such networks the communication is subject to delays or packet losses. When considering feedback communication over real-time networks these communication imperfections must be taken into account for the design and analysis of event-based control methods. The work [98] has investigated event-based state-feedback control under the influence of delays, packet losses and quantized state information. The approaches that have been presented this thesis can be extended in this direction following the design and analysis methods that have been proposed in [98].

A more important issue with respect to imperfect feedback communication is the interplay between the event triggering mechanism and the network protocols which define the rules for the communication over the network. The particular communication protocol can have a severe impact on the behavior of the networked control system, [32], and should be taken into account in the control system design. The performance of a simple event-based controller for systems with integrator dynamics has been analyzed in [38], showing that for some protocols the event-based control is outperformed by a conventional discrete time control. Further research should clarify whether it is possible to avoid undesired effects in the control system by an appropriate design of the event-based controller that is tailored to the respective communication protocol. A promising approach to investigate these systems is in the framework of stochastic hybrid systems, [85], which is expected to yield less conservative results than analysis and design methods that are based on deterministic models of the networked control system.

Improvements of the triggering mechanism. In almost all event-based control methods for interconnected systems, the event triggering mechanism is decentralized in the sense that events are generated based on locally available information only. This approach is reasonable from a networked control perspective, because a centralized event generator that orchestrates the triggering of events in the overall network generally wants to be avoided. However, the decentralized event triggering can lead to the situation where the event conditions in all event generators are satisfied in an arbitrarily small interval which results in a high communication load or even collisions of messages. New triggering mechanisms have to be developed which circumvent these problems by cooperating with each other, aiming at the distribution of events triggered in the overall control system over time, while simultaneously ensuring local performance requirements. A suitable approach to tackle this problem could be to use ideas of periodic event-triggered control, which is a special event-based control concept where the event condition is evaluated periodically [79, 133].

Bibliography

Contributions by the author

[1] M. Sigurani, C. Stöcker, L. Grüne, and J. Lunze. Experimental evaluation of two complementary decentralized event-based control methods. *Control Engineering Practice*, 2014. submitted.

[2] C. Stöcker and J. Lunze. Event-based control of input-output linearizable systems. In *Proc. 18th IFAC World Congress*, pages 10062–10067, 2011.

[3] C. Stöcker and J. Lunze. Event-based control of nonlinear systems: An input-output linearization approach. In *Proc. Joint IEEE Conference on Decision and Control and European Control Conference*, pages 2541–2546, 2011.

[4] C. Stöcker and J. Lunze. Event-based control with incomplete state measurement and guaranteed performance. In *Proc. 3rd IFAC Workshop on Distributed Estimation and Control in Networked Systems*, pages 49–54, 2012.

[5] C. Stöcker and J. Lunze. Input-to-state stability of event-based state-feedback control. In *Proc. 13th European Control Conference*, pages 49–54, 2013.

[6] C. Stöcker and J. Lunze. *Control Theory of Digitally Networked Systems*, chapter Event-based stabilization of interconnected systems, pages 191–203. Springer-Verlag, Heidelberg, 2013.

[7] C. Stöcker and J. Lunze. Event-based feedback control of disturbed input-affine systems. *J. of Appl. Mathematics and Mechanics*, 2013. to be published.

[8] C. Stöcker and J. Lunze. Distributed control of interconnected systems with event-based information requests. In *Proc. 4th IFAC Workshop on Distributed Estimation and Control in Networked Systems*, pages 348–355, 2013.

[9] C. Stöcker and J. Lunze. Distributed event-based control of physically interconnected systems. In *Proc. IEEE Conference on Decision and Control*, pages 7376–7383, 2013.

[10] C. Stöcker, J. Lunze, and C. Ngo. Two methods for the event-based control of interconnected systems and their experimental evaluation. *Automatisierungstechnik*, 60(12): 724–734, 2012.

[11] C. Stöcker, J. Lunze, and D. Vey. Stability analysis of interconnected event-based control loops. In *Proc. IFAC Conference on Analysis and Design of Hybrid Systems*, pages 58–63, 2012.

[12] C. Stöcker, D. Vey, and J. Lunze. Decentralized event-based control: Stability analysis and experimental evaluation. *Nonlinear Analysis: Hybrid Systems*, 10:141–155, 2013.

Supervised theses

[13] M. Bockstegers. Komponentenorientierte Modellierung eines verfahrenstechnischen Prozesses. Diploma thesis, Ruhr-Universität Bochum, Lehrstuhl für Automatisierungstechnik und Prozessinformatik, 2010.

[14] P. Göttel. Entwurf, Analyse und Erprobung einer Methode zur verteilten ereignisbasierten Regelung physikalisch gekoppelter Systeme. Master thesis, Ruhr-Universität Bochum, Lehrstuhl für Automatisierungstechnik und Prozessinformatik, 2014.

[15] D. Hußlein. Analyse und Entwurf einer ereignisbasierten Regelung unter Berücksichtigung von Stellgrößenbeschränkungen. Master thesis, Ruhr-Universität Bochum, Lehrstuhl für Automatisierungstechnik und Prozessinformatik, 2012.

[16] H.C. Ngo Nguyen. Ereignisbasierte Regelung gekoppelter Systeme. Diploma thesis, Ruhr-Universität Bochum, Lehrstuhl für Automatisierungstechnik und Prozessinformatik, 2011.

[17] S. Niemann. Analyse und Erprobung einer Methode zur verteilten ereignisbasierten Regelung mit unvollständiger Zustandsmessung. Bachelor thesis, Ruhr-Universität Bochum, Lehrstuhl für Automatisierungstechnik und Prozessinformatik, 2011.

[18] C. Schymura. Entwurf und Erprobung einer ereignisbasierten Regelung zur Störkompensation für einen thermofluiden Prozess. Master thesis, Ruhr-Universität Bochum, Lehrstuhl für Automatisierungstechnik und Prozessinformatik, 2012.

[19] S. Soub. Erprobung einer Methode zur ereignisbasierten Regelung mit eingeschränktem Netzwerkzugriff. Bachelor thesis, Ruhr-Universität Bochum, Lehrstuhl für Automatisierungstechnik und Prozessinformatik, 2012.

[20] D. Vey. Untersuchung und Erprobung eines Ansatzes zur verteilten ereignisbasierten Regelung schwach gekoppelter Systeme. Master thesis, Ruhr-Universität Bochum, Lehrstuhl für Automatisierungstechnik und Prozessinformatik, 2011.

Further literature

[21] J. Almeida, C. Silvestre, and A.M. Pascoal. Self-triggered state-feedback control of linear plants under bounded distrubances. In *Proc. IEEE Conference on Decision and Control*, pages 7588–7593, 2010.

[22] F.L. Alvardo, J. Meng, C.L. DeMarco, and W.S. Mota. Stability analysis of interconnected power systems coupled with marked dynamics. *IEEE Trans. Power Systems*, 16 (4):695–701, 2001.

[23] A. Anta and P. Tabuada. To sample or not to sample: Self-triggered control for nonlinear systems. *IEEE Trans. Autom. Control*, 55(9):2030–2042, 2010.

[24] A. Anta and P. Tabuada. On the minimum attention and anytime attention problems for nonlinear systems. In *Proc. IEEE Conference on Decision and Control*, pages 3234–3239, 2010.

[25] A. Anta and P. Tabuada. Exploiting isochrony in self-triggered control. *IEEE Trans. Autom. Control*, 57(4):950–962, 2012.

[26] D. Antunes, W.P.M.H. Heemels, and P. Tabuada. Dynamic programming formulation of periodic event-triggered control: Performance guarantees and co-design. In *Proc. IEEE Conference on Decision and Control*, pages 7212–7217, 2012.

[27] J. Araújo, A. Anta, M. Mazo Jr., J. Faria, A. Hernandez, P. Tabuada, and K.H. Johansson. Self-triggered control over wireless sensor and actuator systems. In *Proc. IEEE Conference on Distributed Computing in Sensor Systems*, pages 1–9, 2011.

[28] K. Arzén. A simple event-based PID controller. In *Proc. 14th IFAC World Congress*, pages 423–428, 1999.

[29] K.J. Åström. Event based control. In *Analysis and Design of Nonlinear Control Systems: In Honor of Alberto Isidori*. Springer Verlag, 2007.

[30] K.J. Åström and B.M. Bernhardsson. Comparison of periodic and event-based sampling for first-order stochastic systems. In *Proc. 14th IFAC World Congress*, pages 301–306, 1999.

[31] K.J. Åström and B.M. Bernhardsson. Comparison of Riemann and Lebesgue sampling for first order stochastic systems. In *Proc. IEEE Conference on Decision and Control*, pages 2011–2016, 2002.

[32] J. Baillieul and P.J. Antsaklis. Control and communication challanges in networked real-time systems. *Proceedings of the IEEE*, 95(1):9–28, 2007.

[33] G.A. Bekey and R. Tomovic. Sensitivity of discrete systems to variation of sampling interval. *IEEE Trans. Autom. Control*, 11(2):284–287, 1966.

[34] J.D.J. Barradas Berglind, T.M.P. Gommans, and W.P.M.H. Heemels. Self-triggered MPC for constrained linear systems and quadratic costs. In *Proc. IFAC Nonlinear Model Predictive Control Conference*, pages 342–348, 2012.

[35] D. Bernardini and A. Bemporad. Energy-aware robust model predictive control based on noisy wireless sensors. *Automatica*, 48(1):36–44, 2012.

[36] D.S. Bernstein. *Matrix Mathematics: Theory, Facts and Formulas*. Princton University Press, 2009.

[37] M. Beschi, S. Dormido, J. Sanchez, and A. Visioli. An automatic tuning procedure for an event-based PI controller. In *Proc. IEEE Conference on Decision and Control*, pages 7437–7442, 2013.

[38] R. Blind and F. Allgöwer. On time-triggered and event-based control of integrator systems over a shared communication system. *Mathematics of Control, Signals, and Systems*, 25(4):517–557, 2013.

[39] D.P. Borgers and W.P.M.H. Heemels. On minimum inter-event times in event-triggered control. In *Proc. IEEE Conference on Decision and Control*, pages 7370–7375, 2013.

[40] R.W. Brockett. Minimum attention control. In *Proc. IEEE Conference on Decision and Control*, pages 2628–2632, 1997.

[41] A. Cervin and T. Henningsson. Scheduling of event-triggered controllers on a shared network. In *Proc. IEEE Conference on Decision and Control*, pages 3601–3606, 2008.

[42] J. Chacón, J. Sánchez, L. Yebra, A. Visioli, and S. Dormido. Experimental study of two event-based PI controllers in a solar distributed collector field. In *Proc. European Control Conference*, pages 626–631, 2013.

[43] D. Ciscato and L. Mariani. On increasing sampling efficiency by adaptive sampling. *IEEE Trans. Autom. Control*, 12(3):318, 1967.

[44] R. Cogill. Event-based control using quadratic approximate value functions. In *Proc. Joint IEEE Conference on Decision and Control and Chinese Control Conference*, pages 5883–5888, 2009.

[45] R. Cogill, S. Lall, and J.P. Hespanha. A constant factor approximation algorithm for event-based sampling. In *Proc. American Control Conference*, pages 305–311, 2007.

[46] N. Cordoso de Castro, D.E. Quevedo, F. Garin, and C. Canudas de Wit. Smart energy-aware sensors for event-based control. In *Proc. IEEE Conference on Decision and Control*, pages 7224–7229, 2012.

[47] C. De Persis, R. Sailer, and F. Wirth. Parsimonious event-triggered distributed control: A Zeno free approach. *Automatica*, 49(7):2116–2124, 2013.

[48] C. De Persis, R. Sailer, and F. Wirth. On inter-sampling times for event-triggered large-scale linear systems. In *Proc. IEEE Conference on Decision and Control*, pages 5301–5306, 2013.

[49] O. Demir and J. Lunze. Cooperative control of multi-agent systems with event-based communication. In *Proc. American Control Conference*, pages 4504–4509, 2012.

[50] O. Demir and J. Lunze. Event-based synchronisation of multi-agent systems. In *Proc. IFAC Conference on Analysis and Design of Hybrid Systems*, pages 1–6, 2012.

[51] B. Demirel, V. Gupta, and M. Johansson. On the trade-off between control performance and communication cost for event-triggered control over lossy networks. In *Proc. European Control Conference*, pages 1168–1174, 2013.

[52] D.V. Dimarogonas, E. Frazzoli, and K.H. Johansson. Distributed event-triggered control for multi-agent systems. *IEEE Trans. Autom. Control*, 57(5):1291–1297, 2012.

[53] M.C.F. Donkers. *Networked and Event-Triggered Control Systems*. PhD thesis, Eindhoven University of Technology, 2011.

[54] M.C.F. Donkers and W.P.M.H. Heemels. Output-based event-triggered control with guaranteed $\mathcal{L}_\infty$-gain and improved and decentralised event-triggering. *IEEE Trans. Autom. Control*, 57(6):1362–1376, 2012.

[55] M.C.F. Donkers, P. Tabuada, and W.P.M.H. Heemels. On the minimum attention control problem for linear systems: A linear programming approach. In *Proc. IEEE Conference on Decision and Control*, pages 4717–4722, 2011.

[56] A. Eqtami, D.V. Dimarogonas, and K.J. Kyriakopoulos. Aperiodic model predictive control via perturbation analysis. In *Proc. IEEE Conference on Decision and Control*, pages 7193–7198, 2012.

[57] A. Eqtami, D.V. Dimarogonas, and K.J. Kyriakopoulos. Event-based model predictive control for the cooperation of distributed agents. In *Proc. American Control Conference*, pages 6473–6478, 2012.

[58] T. Estrada, H. Lin, and P.J. Antsaklis. Model-based control with intermittent feedback. In *Proc. Mediterranean Conference on Control and Automation*, pages 1–6, 2006.

[59] Y. Fan, G. Feng, Y. Wang, and C. Song. Distributed event-triggered control of multi-agent systems with combinational measurements. *Automatica*, 49(2):671–675, 2013.

[60] C. Fiter, L. Hetel, W. Perruquetti, and J.-P. Richard. A state dependent sampling for linear state feedback. *Automatica*, 48(8):1860–1867, 2012.

[61] E. Garcia and P.J. Antsaklis. Decentralized model-based event-triggered control of networked systems. In *Proc. American Control Conference*, pages 6485–6490, 2012.

[62] E. Garcia and P.J. Antsaklis. Output feedback model-based control of uncertain discrete-time systems with network induced delays. In *Proc. IEEE Conference on Decision and Control*, pages 6647–6652, 2012.

[63] E. Garcia and P.J. Antsaklis. Model-based event-triggered control for systems with quantization and time-varying network delays. *IEEE Trans. Autom. Control*, 58(2):422–434, 2013.

[64] E. Garcia, Y. Cao, H. Yu, P. Antsaklis, and D. Casbeer. Decentralised event-triggered cooperative control with limited communication. *Int. J. Control*, 86(9):1479–1488, 2013.

[65] K. Gatsis, M. Pajic, A. Ribeiro, and G.J. Pappas. Power-aware communication for wireless sensor-actuator systems. In *Proc. IEEE Conference on Decision and Control*, pages 4006–4011, 2013.

[66] P.J. Gawthrop and L.B. Wang. Event-driven intermittent control. *Int. J. Control*, 82(12): 2235–2248, 2009.

[67] R. Goebel, R.G. Sanfelice, and A.R. Teel. Hybrid dynamical systems. *IEEE Control Systems Magazine*, 29(2):28–93, 2009.

[68] L. Grüne, S. Jerg, O. Junge, D. Lehmann, J. Lunze, F. Müller, and M. Post. Two complementary approaches to event-based control. *Automatisierungstechnik*, 58(4):173–182, 2010.

[69] L. Grüne and F. Müller. An algorithm for event-based optimal feedback control. In *Proc. IEEE Confeference on Decision and Control*, pages 5311–5316, 2009.

[70] M. Guinaldo, D.V. Dimarogonas, K.H. Johansson, J. Sánchez, and S. Dormido. Distributed event-based control for interconnected linear systems. In *Proc. IEEE Conference on Decision and Control*, pages 2553–2558, 2011.

[71] M. Guinaldo, D. Lehmann, J. Sánchez, S. Dormido, and K.H. Johansson. Distributed event-triggered control with network delays and packet losses. In *Proc. IEEE Conference on Decision and Control*, pages 1–6, 2012.

[72] Y. Guo, D.J. Hill, and Y. Wang. Nonlinear decentralized control of large-scale power systems. *Automatica*, 36:1275–1289, 2000.

[73] S.C. Gupta. Increasing the sampling efficiency for a control system. *IEEE Trans. Autom. Control*, 8(3):263–264, 1963.

[74] D. Han, Y. Mo, J. Wu, B. Sinopoli, and L. Shi. Stochastic event-triggered sensor scheduling for remote state estimation. In *Proc. IEEE Conference on Decision and Control*, pages 6079–6084, 2013.

[75] W.P.M.H. Heemels and M.C.F. Donkers. Model-based periodic event-triggered control for linear systems. *Automatica*, 49(3):698–711, 2013.

[76] W.P.M.H. Heemels, R.J.A. Gorter, A. van Zijl, P.P.J. van den Bosch, S. Weiland, W.H.A. Hendrix, and M.R. Vonder. Asynchronous measurement and control: a case study on motor synchronization. *Contr. Eng. Practice*, 7(12):1467–1482, 1999.

[77] W.P.M.H. Heemels, J.H. Sandee, and P.P.J. van den Bosch. Analysis of event-driven controllers for linear systems. *Int. J. Control*, 81(4):571–590, 2008.

[78] W.P.M.H. Heemels, K.H. Johansson, and P. Tabuada. An introduction to event-triggered and self-triggered control. In *Proc. IEEE Conference on Decision and Control*, pages 3270–3285, 2012.

[79] W.P.M.H. Heemels, M.C.F. Donkers, and A.R. Teel. Periodic event-triggered control for linear systems. *IEEE Trans. Autom. Control*, 58(4):847–861, 2013.

[80] E. Hendricks, M. Jensen, A. Chevalier, and T. Vesterholm. Problems in event based engine control. In *Proc. American Control Conference*, pages 1585–1587, 1994.

[81] T. Henningsson and A. Cervin. Comparison of LTI and event-based control for a moving cart with quantized position measurement. In *Proc. European Control Conference*, pages 3791–3796, 2009.

[82] T. Henningsson and A. Cervin. A simple model for the interface between event-based control loops using a shared medium. In *Proc. IEEE Conference on Decision and Control*, pages 3240–3245, 2010.

[83] T. Henningsson, E. Johannesson, and A. Cervin. Sporadic event-based control of first-order linear stochastic systems. *Automatica*, 44(11):2890–2895, 2008.

[84] E. Henriksson, D.E. Quevedo, H. Sandberg, and K.H. Johansson. Self-triggered model predictive control for network scheduling and control. In *Proc. of International Symposium on Advanced Control of Chemical Processes*, pages 432–438, 2012.

[85] J.P. Hespanha. Modeling and analysis of stochastic hybrid systems. *IEE Proc. — Control Theory & Applications,* Special Issue on Hybrid Systems, 153(5):520–535, 2007.

[86] J.P. Hespanha, P. Naghshtabrizi, and Y. Xu. A survey of recent results in networked control systems. *Proceedings of the IEEE*, 95(1):138–162, 2007.

[87] D. Hristu-Varsakelis and P.R. Kumar. Interrupt-based feedback control over a shared communication medium. In *Proc. IEEE Conference on Decision and Control*, pages 3223–3228, 2002.

[88] T.C. Hsia. A unified approach to adaptive sampling system design. In *Proc. IEEE Conference on Decision and Control and Symposium on Adaptive Processes*, pages 618–622, 1972.

[89] M. Jelali. Performance assessment of control systems in rolling mills - Application to strip thickness and flatness control. *Journal of Process Control*, 17(10):805–816, 2007.

[90] M. Johansson. *Piecewise Linear Control Systems: A Computational Approach.* Springer Verlag, 2003.

[91] Y. Kadowaki and H. Ishii. Event-based distributed clock synchronization for wireless sensor networks. In *Proc. IEEE Conference on Decision and Control*, pages 6747–6752, 2013.

[92] P.L. Kempker, A.C.M. Ran, and J.H. van Schuppen. A linear-quadratic coordination control procedure with event-based communication. In *Proc. IFAC Conference on Analysis and Design of Hybrid Systems*, pages 13–18, 2012.

[93] H.K. Khalil. *Nonlinear Systems*. Prentice Hall, New Jersey, 2002.

[94] K.-D. Kim and P.R. Kumar. Cyber-physical systems: A perspective at the centennial. *Proceedings of the IEEE*, 100:1287–1308, 2012.

[95] E. Kofman and J.H. Braslavsky. Level crossing sampling in feedback stabilization under data-rate constraints. In *Proc. IEEE Conference on Decision and Control*, pages 4423–4428, 2006.

[96] W.H. Kwon, Y.H. Kim, S.J. Lee, and K. Paek. Event-based modeling and control for the burnthrough point in sintering processes. *IEEE Trans. Control Syst. Technol.*, 7(1): 31–41, 1999.

[97] S. Lee, W. Liu, and I. Hwang. Event-based state estimation algorith using Markov chain approximation. In *Proc. IEEE Conference on Decision and Control*, pages 6998–7003, 2013.

[98] D. Lehmann. *Event-Based State-Feedback Control*. Logos-Verlag, Berlin, 2011.

[99] D. Lehmann and J. Lunze. Event-based control using quantized state information. In *Proc. IFAC Workshop on Distributed Estimation and Control in Networked Systems*, pages 1–6, 2010.

[100] D. Lehmann and J. Lunze. Extension and experimental evaluation of an event-based state-feedback approach. *Contr. Eng. Practice*, 19(2):101–112, 2011.

[101] D. Lehmann and J. Lunze. Event-based output-feedback control. In *Proc. Mediterranean Conference on Control and Automation*, pages 982–987, 2011.

[102] D. Lehmann and J. Lunze. Event-based control with communication delays and packet losses. *Int. J. Control*, 85(5):563–577, 2012.

[103] D. Lehmann, G.A. Kiener, and K.H. Johansson. Event-triggered PI control: Saturating actuators and anti-windup compensation. In *Proc. IEEE Conference on Decision and Control*, pages 6566–6571, 2012.

[104] D. Lehmann, J. Lunze, and K.H. Johansson. Comparison between sampled-data control, deadband control and model-based event-triggered control. In *Proc. IFAC Conference on Analysis and Design of Hybrid Systems*, pages 7–12, 2012.

[105] D. Lehmann, E. Henriksson, and K.H. Johansson. Event-triggered model predictive control of discrete-time linear systems subject to disturbances. In *Proc. European Control Conference*, pages 1156–1161, 2013.

[106] M.D. Lemmon. Event-triggered feedback in control, estimation, and optimization. In A. Bemporad, W.P.M.H. Heemels, and M. Johansson, editors, *Networked Control Systems*, volume 405 of *Lecture Notes in Control and Information Sciences*, pages 293–358. Springer-Verlag, Berlin Heidelberg, 2010.

[107] L. Li and M.D. Lemmon. Weakly coupled event triggered output feedback control in wireless networked control systems. In *Allerton Conference on Communication, Control and Computing*, Urbana-Champaign, USA, 2011.

[108] L. Li, B. Hu, and M.D. Lemmon. Resilient event triggered systems with limited communication. In *Proc. IEEE Conference on Decision and Control*, pages 6577–6582, 2012.

[109] L. Li, X. Wand, and M.D. Lemmon. Stabilizing bit-rate of disturbed event triggered control systems. In *Proc. IFAC Conference on Analysis and Design of Hybrid Systems*, pages 70–75, 2012.

[110] D. Liberzon and S. Trenn. The bang-bang funnel controller. In *Proc. IEEE Conference on Decision and Control*, pages 690–695, 2010.

[111] T. Liu, D.J. Hill, and B. Liu. Synchronization of dynamical networks with distributed event-based communication. In *Proc. IEEE Conference on Decision and Control*, pages 7199–7204, 2012.

[112] D. G. Luenberger. *Linear and Nonlinear Programming*. Addison Wesley, 1984.

[113] J. Lunze. *Feedback Control of Large-Scale Systems*. Prentice Hall, London, 1992.

[114] J. Lunze and F. Lamnabhi-Lagarrigue, editors. *Handbook of Hybrid Systems Control - Theory, Tools, Applications*. Cambridge University Press, Cambridge, 2009.

[115] J. Lunze and D. Lehmann. A state-feedback approach to event-based control. *Automatica*, 46(1):211–215, 2010.

[116] M.H. Mamduhi, A. Molin, and S. Hirche. Stability analysis of stochastic prioritized dynamic scheduling for ressource-aware heterogeneous multi-loop control systems. In *Proc. IEEE Conference on Decision and Control*, pages 7390–7396, 2013.

[117] M. Mazo Jr. and M. Cao. Decentralized event-triggered control with asynchronous updates. In *Joint IEEE Conference on Decision and Control and European Control Conference*, pages 2547–2552, 2011.

[118] M. Mazo Jr. and P. Tabuada. Input-to-state stability of self-triggered control systems. In *Porc. Joint IEEE Conference on Decision and Control and Chinese Control Conference*, pages 928–933, 2009.

[119] M. Mazo Jr. and P. Tabuada. Towards decentralized event-triggered implementations of centralized control laws. In *Proc. Int. Workshop on Networks of Cooperating Objects*, 2010.

[120] M. Mazo Jr. and P. Tabuada. Decentralized event-triggered control over wireless sensor/actuator networks. *IEEE Trans. Autom. Control*, 56(10):2456–2461, 2011.

[121] M. Mazo Jr., A. Anta, and P. Tabuada. An ISS self-triggered implementation of linear controllers. *Automatica*, 46(8):1310–1314, 2010.

[122] X. Meng and T. Chen. Event based agreement protocols for multi-agent networks. *Automatica*, 49(7):2125–2132, 2013.

[123] M Miskowicz. Send-on-delta concept: An event-based data reporting strategy. *Sensors*, 6:49–63, 2006.

[124] J.R. Mitchell and W.L. McDaniel Jr. Adaptive sampling technique. *IEEE Trans. Autom. Control*, 14(2):200–201, 1969.

[125] A. Molin and S. Hirche. Structural characterization of optimal event-based controllers for linear stochastic systems. In *Proc. IEEE Conference on Decision and Control*, pages 3227–3233, 2010.

[126] A. Molin and S. Hirche. Adaptive event-triggered control over a shared network. In *Proc. IEEE Conference on Decision and Control*, pages 6591–6596, 2012.

[127] A. Molin and S. Hirche. An iterative algorithm for optimal event-triggered estimation. In *Proc. IFAC Conference on Analysis and Design of Hybrid Systems*, pages 64–69, 2012.

[128] A. Molin and S. Hirche. On the optimality of certainty equivalence for event-triggered control systems. *IEEE Trans. Autom. Control*, 58(2):470–474, 2013.

[129] L.A. Montestruque and P.J. Antsaklis. Stability of model-based networked control systems with time-varying transmission times. *IEEE Trans. Autom. Control*, 49(9):1562–1572, 2004.

[130] C. Nowzari and J. Cortés. Self-triggered coordination of robotic networks for optimal deployment. *Automatica*, 48(6):1077–1087, 2012.

[131] P.G. Otanez, J.R. Moyne, and D.M. Tilbury. Using deadbands to reduce communication in networked control systems. In *Proc. American Control Conference*, 2002.

[132] R. Postoyan, P. Tabuada, D. Nešić, and A. Anta. Event-triggered and self-triggered stabilization of distributed networked control systems. In *Proc. Joint IEEE Conference on Decision and Control and European Control Conference*, pages 2565–2570, 2011.

[133] R. Postoyan, A. Anta, W.P.M.H. Heemels, P. Tabuada, and D. Nešić. Periodic event-triggered control for nonlinear sytems. In *Proc. IEEE Conference on Decision and Control*, pages 7397–7402, 2013.

[134] M. Rabi and J.S. Baras. Level-triggered control of a scalar linear system. In *Proc. Mediterranean Conference on Control and Automation*, pages 1–6, 2007.

[135] M. Rabi and K.H. Johansson. Scheduling packets for event-triggered control. In *Proc. European Control Conference*, pages 3779–3784, 2009.

[136] M. Rabi, G.V. Moustakides, and J.S. Baras. Adaptive sampling for linear state-estimation. *SIAM J. Control Optim.*, 50(2):672–702, 2012.

[137] C. Ramesh, H. Sandberg, L. Bao, and K.H. Johansson. On the dual effect in state-based scheduling of networked control systems. In *Proc. American Control Conference*, pages 2216–2221, 2011.

[138] J.H. Sandee, W.P.M.H. Heemels, S.B.F. Hulsenboom, and P.P.J. van den Bosch. Analysis and experimental validation of a sensor-based event-driven controller. In *Proc. American Control Conference*, pages 2867–2874, 2007.

[139] G. Seyboth, D.V. Dimarogonas, and K.H. Johansson. Event-based broadcasting for multi-agent average consensus. *Automatica*, 49(1):245–252, 2013.

[140] J. Sijs, M. Lazar, and W.P.M.H. Heemels. On integration of event-based estimation and robust MPC in a feedback loop. In *Hybrid Systems: Computation and Control*, pages 31–40, 2010.

[141] E.D. Sontag. Smooth stabilization implies coprime factorization. *IEEE Trans. Autom. Control*, 34:435–443, 1989.

[142] E.D. Sontag and Y. Wang. New characterizations of input-to-state stability. *IEEE Trans. Autom. Control*, 41(9):1283–1294, 1996.

[143] A. Swarnakar, H.J. Marquez, and T. Chen. Robust stabilization of nonlinear interconnected systems with application to an industrial utility boiler. *Contr. Eng. Practice*, 15: 639–654, 2007.

[144] P. Tabuada. Event-triggered real-time scheduling of stabilizing control tasks. *IEEE Trans. Autom. Control*, 52(9):1680–1685, 2007.

[145] P. Tallapragada and N. Chopra. On event triggered trajectory tracking for control affine nonlinear systems. In *Proc. Joint IEEE Conference on Decision and Control and European Control Conference*, pages 5377–5382, 2011.

[146] P. Tallapragada and N. Chopra. Event-triggered output feedback control for LTI systems. In *Proc. IEEE Conference on Decision and Control*, pages 6597–6602, 2012.

[147] P. Tallapragada and N. Chopra. On co-design of event trigger and quantizer for emulation based control. In *Proc. American Control Conference*, pages 3772–3777, 2012.

[148] P. Tallapragada and N. Chopra. Event-triggered decentralized dynamic output feedback control for LTI systems. In *Proc. IFAC Workshop on Estimation and Control of Networked Systems*, pages 31–36, 2012.

[149] P. Tallapragada and N. Chopra. On event triggered tracking for nonlinear systems. *IEEE Trans. Autom. Control*, 58(9):2343–2348, 2013.

[150] P. Tallapragada and N. Chopra. Event-triggered dynamic output feedback control of LTI systems over sensor-controller-actuator networks. In *Proc. IEEE Conference on Decision and Control*, pages 4625–4630, 2013.

[151] A.S. Tanenbaum and D.J. Wetherall. *Computer Networks*. Prentice Hall, 2010.

[152] U. Tiberi and K.H. Johansson. A simple self-triggered sampler for nonlinear systems. In *Proc. IFAC Conference on Analysis and Design of Hybrid Systems*, pages 76–81, 2012.

[153] U. Tiberi, C. Fischione, K.H. Johansson, and M.D. Di Benedetto. Adaptive self-triggered control over IEEE 802.15.4 networks. In *Proc. IEEE Conference on Decision and Control*, pages 2099–2104, 2010.

[154] Y. Tipsuwan and M.-Y. Chow. Control methodologies in networked control systems. *Contr. Eng. Practice*, 11(10):1099–1111, 2003.

[155] S. Trimpe and R. D'Andrea. An experimental demonstration of a distributed and event-based state estimation algorithm. In *Proc. 18th IFAC World Congress*, pages 8811–8818, 2011.

[156] S. Trimpe and R. D'Andrea. Event-based state estimation with variance-based triggering. In *Proc. IEEE Conference on Decision and Control*, pages 6583–6590, 2012.

[157] P. Varutti, B. Kern, T. Faulwasser, and R. Findeisen. Event-based model predictive control for networked control systems. In *Proc. Joint IEEE Conference on Decision and Control and Chinese Control Conference*, pages 567–572, 2009.

[158] P. Varutti, T. Faulwasser, B. Kern, M. Kögel, and R. Findeisen. Event-based reduced-attention predictive control for nonlinear uncertain systems. In *Proc. IEEE Multi-Conference on Systems and Control*, pages 1085–1090, 2010.

[159] M. Velasco, P. Martí, and J.M. Fuertes. The self-triggered task model for real-time control systems. In *Proc. IEEE Real-Time Systems Symposium*, pages 67–70, 2003.

[160] J.L.C. Verhaegh, T.M.P. Gommans, and W.P.M.H. Heemels. Extension and evaluation of model-based periodic event-triggered control. In *Proc. European Control Conference*, pages 1138–1144, 2013.

[161] X. Wang and M.D. Lemmon. Event-triggered broadcasting across distributed networked control systems. In *Proc. American Control Conference*, pages 3139–3144, 2008.

[162] X. Wang and M.D. Lemmon. Self-triggered feedback control systems with finite-gain $\mathcal{L}_2$ stability. *IEEE Trans. Autom. Control*, 45(3):452–467, 2009.

[163] X. Wang and M.D. Lemmon. Event-triggering in distributed networked systems with data dropouts and delays. In R. Majumdar and P. Tabuada, editors, *Hybrid Systems: Computation and Control*, pages 366–380, 2009.

[164] X. Wang and M.D. Lemmon. Self-triggering under state-independent disturbances. *IEEE Trans. Autom. Control*, 55(6):1494–1500, 2010.

[165] X. Wang and M.D. Lemmon. Event-triggering in distributed networked control systems. *IEEE Trans. Autom. Control*, 56(3):586–601, 2011.

[166] X. Wang and M.D. Lemmon. On event design in event-triggered feedback systems. *Automatica*, 47(10):2319–2322, 2012.

[167] J. Weimer, J. Araújo, and K.H. Johansson. Distributed event-triggered estimation in networked systems. In *Proc. IFAC Conference on Analysis and Design of Hybrid Systems*, pages 178–185, 2012.

[168] K. Xin, P. Cheng, J. Chen, and L. Xie. Sensor data forwarding strategies for state estimation in multi-hop wireless networks. In *Proc. IEEE Conference on Decision and Control*, pages 4772–4777, 2013.

[169] Y. Xu and J.P. Hespanha. Communication logics for networked control systems. In *Proc. American Control Conference*, pages 572–577, 2004.

[170] Y. Xu and J.P. Hespanha. Optimal communication logics in networked control systems. In *Proc. IEEE Conference on Decision and Control*, pages 3527–3532, 2004.

[171] D. Xue and S. Hirche. Event-triggered consensus of heterogeneous multi-agent systems with double-integrator dynamics. In *Proc. European Control Conference*, pages 1162–1167, 2013.

[172] J.K. Yook, D.M. Tilbury, and N.R. Soparkar. Trading computation for bandwidth: Reducing communication in distributed control systems using state estimators. *IEEE Trans. Control Syst. Technol.*, 10(4):503–518, 2002.

[173] H. Yu and P.J. Antsaklis. Formation control of multi-agent systems with connectivity preservation by using both event-driven and time-driven communication. In *Proc. IEEE Conference on Decision and Control*, pages 7218–7223, 2012.

[174] H. Yu and P.J. Antsaklis. Event-triggered output feedback control for networked control systems using passivity: Achieving $\mathcal{L}_2$ stability in the presence of communication delays and signal quantization. *Automatica*, 49(1):30–38, 2013.

Appendix

A Thermofluid process models

This appendix presents the nonlinear model of the thermofluid process described in Sec. 2.4 as well as the linearized model that is applied for the design and analysis of the event-based controllers according to the proposed approaches. This model can be found in [1].

A.1 Nonlinear model

The continuous flow process is represented by the nonlinear model

$$\dot{l}_{\mathrm{B}}(t) = A_{\mathrm{B}}^{-1}\Big(q_{1\mathrm{B}}(u_{\mathrm{T1}}(t)) + q_{\mathrm{SB}}(l_{\mathrm{S}}(t), u_{\mathrm{SB}}) - q_{\mathrm{BW}}(l_{\mathrm{B}}(t), u_{\mathrm{BW}}) - q_{\mathrm{BS}}(l_{\mathrm{B}}(t), u_{\mathrm{BS}})\Big) \tag{A.1a}$$

$$\begin{aligned}\dot{\vartheta}_{\mathrm{B}}(t) = (A_{\mathrm{B}} l_{\mathrm{B}}(t))^{-1}\Big(& q_{1\mathrm{B}}(u_{\mathrm{T1}}(t))(\vartheta_1 - \vartheta_{\mathrm{B}}(t)) + q_{\mathrm{SB}}(l_{\mathrm{S}}(t), u_{\mathrm{SB}})(\vartheta_{\mathrm{S}}(t) - \vartheta_{\mathrm{B}}(t)) \\ &+ q_{\mathrm{C}}(u_{\mathrm{CU}}(t))(\vartheta_{\mathrm{C}} - \vartheta_{\mathrm{B}}(t)) + H_{\mathrm{B}} d_{\mathrm{H}}(t)\Big)\end{aligned} \tag{A.1b}$$

$$\begin{aligned}\dot{l}_{\mathrm{S}}(t) = A_{\mathrm{S}}^{-1}\Big(& q_{3\mathrm{S}}(u_{\mathrm{T3}}(t)) + q_{\mathrm{BS}}(l_{\mathrm{B}}(t), u_{\mathrm{BS}}) - q_{\mathrm{SW}}(l_{\mathrm{S}}(t), u_{\mathrm{SW}}) - q_{\mathrm{SB}}(l_{\mathrm{S}}(t), u_{\mathrm{SB}}) \\ &+ q_{\mathrm{FS}}(d_{\mathrm{F}}(t))\Big)\end{aligned} \tag{A.1c}$$

$$\begin{aligned}\dot{\vartheta}_{\mathrm{S}}(t) = (A_{\mathrm{S}} l_{\mathrm{S}}(t))^{-1}\Big(& q_{3\mathrm{S}}(u_{\mathrm{T3}}(t))(\vartheta_3 - \vartheta_{\mathrm{S}}(t)) + q_{\mathrm{BS}}(l_{\mathrm{B}}(t), u_{\mathrm{BS}})(\vartheta_{\mathrm{B}}(t) - \vartheta_{\mathrm{S}}(t)) \\ &+ q_{\mathrm{FS}}(d_{\mathrm{F}}(t))(\vartheta_{\mathrm{F}} - \vartheta_{\mathrm{S}}(t)) + H_{\mathrm{S}} u_{\mathrm{H}}(t)\Big).\end{aligned} \tag{A.1d}$$

Here,

$$q_{1\mathrm{B}}(u_{\mathrm{T1}}(t)) = 1.61 \times 10^{-4} \cdot u_{\mathrm{T1}}(t) \tag{A.2a}$$

$$q_{3\mathrm{S}}(u_{\mathrm{T3}}(t)) = 1.81 \times 10^{-4} \cdot u_{\mathrm{T3}}(t) \tag{A.2b}$$

denote the flows from the storage tanks T_1 and T_3 to the reactors TB and TS, respectively.

$$q_{\mathrm{C}}(u_{\mathrm{CU}}(t)) = 0.97 \times 10^{-4} \cdot u_{\mathrm{CU}}(t) \tag{A.2c}$$

is the flow of the coolant and

$$q_{\mathrm{BS}}(l_{\mathrm{B}}(t), u_{\mathrm{BS}}) = K_{\mathrm{BS}}(u_{\mathrm{BS}})\sqrt{2gl_{\mathrm{B}}(t)} \tag{A.2d}$$
$$K_{\mathrm{BS}}(u_{\mathrm{BS}}) = 10^{-4} \cdot \begin{cases} 1.02 \cdot u_{\mathrm{BS}}, & 0 \leq u_{\mathrm{BS}} \leq 0.1 \\ 2.13 \cdot u_{\mathrm{BS}} - 0.11, & 0.1 < u_{\mathrm{BS}} \leq 1 \end{cases}$$

$$q_{\mathrm{SB}}(l_{\mathrm{S}}(t), u_{\mathrm{SB}}) = K_{\mathrm{SB}}(u_{\mathrm{SB}})\sqrt{2gl_{\mathrm{S}}(t)} \tag{A.2e}$$
$$K_{\mathrm{SB}}(u_{\mathrm{SB}}) = 10^{-4} \cdot \begin{cases} 0.90 \cdot u_{\mathrm{SB}}, & 0 \leq u_{\mathrm{SB}} \leq 0.1 \\ 1.68 \cdot u_{\mathrm{SB}} - 0.08, & 0.1 < u_{\mathrm{SB}} \leq 1 \end{cases}$$

denote the flows from reactor TB to reactor TS and vice versa with the specific valve parameters K_{BS} and K_{SB} (m^3/m). Finally,

$$q_{\mathrm{BW}}(l_{\mathrm{B}}(t), u_{\mathrm{BW}}) = K_{\mathrm{BW}}(u_{\mathrm{TB}})\sqrt{2gl_{\mathrm{B}}(t)} \tag{A.2f}$$
$$K_{\mathrm{BW}}(u_{\mathrm{BW}}) = 10^{-4} \cdot \begin{cases} 0.96 \cdot u_{\mathrm{TB}}, & 0 \leq u_{\mathrm{BW}} \leq 0.1 \\ 2.01 \cdot u_{\mathrm{TB}} - 0.10, & 0.1 < u_{\mathrm{BW}} \leq 1 \end{cases}$$
$$q_{\mathrm{SW}}(l_{\mathrm{S}}(t), u_{\mathrm{SW}}) = K_{\mathrm{SW}}(u_{\mathrm{SW}})\sqrt{2gl_{\mathrm{S}}(t)} \tag{A.2g}$$
$$K_{\mathrm{SW}}(u_{\mathrm{SW}}) = 10^{-4} \cdot \begin{cases} 0.79 \cdot u_{\mathrm{SW}}, & 0 \leq u_{\mathrm{SW}} \leq 0.1 \\ 1.42 \cdot u_{\mathrm{SW}} - 0.06, & 0.1 < u_{\mathrm{SW}} \leq 1 \end{cases}$$

denote flows of volume from the reactors TB and TS into the buffer reactor TW with the specific valve parameters K_{BW} and K_{SW} (m^3/m). All flows have the unit m^3/s. All parameters are listed in Table A.1.

Table A.1: Parameters

Parameter	Value	Meaning
A_{B}	$0.07\ \mathrm{m}^2$	Cross sectional area of reactor TB
A_{S}	$0.07\ \mathrm{m}^2$	Cross sectional area of reactor TS
g	$9.81\ \mathrm{m/s}^2$	Gravitation constant
H_{B}	$10^{-3} \cdot 4.8\ \mathrm{m}^3\mathrm{K/s}$	Heat coefficient of the heating in reactor TB
H_{S}	$10^{-3} \cdot 0.8\ \mathrm{m}^3\mathrm{K/s}$	Heat coefficient of the heating in reactor TS
ϑ_1	$294.15\ \mathrm{K}$	Temperature of the fluid in tank T_1
ϑ_3	$294.15\ \mathrm{K}$	Temperature of the fluid in tank T_3
ϑ_{C}	$282.65\ \mathrm{K}$	Temperature of the coolant
ϑ_{F}	$294.15\ \mathrm{K}$	Temperature of the water supply

Due to technical limitations the levels $l_{\mathrm{B}}(t)$ and $l_{\mathrm{S}}(t)$, as well as the temperatures $\vartheta_{\mathrm{B}}(t)$ and $\vartheta_{\mathrm{S}}(t)$ are restricted to

$$l_{\mathrm{B}} \in [0.26; 0.40]\ \mathrm{m}, \quad \vartheta_{\mathrm{B}} \in [285.65; 323.15]\ \mathrm{K} \tag{A.3a}$$

$$l_{\mathrm{S}} \in [0.26; 0.40]\ \mathrm{m}, \quad \vartheta_{\mathrm{S}} \in [293.15; 323.15]\ \mathrm{K}. \tag{A.3b}$$

The control inputs u_{T1}, u_{CU}, u_{T3} and u_{H} are limited to variations in the range

$$u_{\mathrm{T1}} \in [0; 1], \quad u_{\mathrm{CU}} \in [0; 1], \quad u_{\mathrm{T3}} \in [0; 1], \quad u_{\mathrm{H}} \in [0; 1].$$

The components which are used for control are highlighted in gray in Fig. 2.4. The disturbance characteristics are accomplished by means of the heating with disturbance input $d_{\mathrm{H}}(t)$ in reactor TB and the additional water inflow in reactor TS that is set by the valve angle $d_{\mathrm{F}}(t)$. The disturbances are considered to be bounded to

$$d_{\mathrm{H}} \in \mathcal{D}_1 = [0; 0.2], \quad d_{\mathrm{F}} \in \mathcal{D}_2 = [0; 0.3]. \tag{A.4}$$

A.2 Linearized models

As the event-based control approaches proposed in this thesis rely on linear process models, the following two paragraphs present a linearized model of the nonlinear model (A.1), (A.2f) with two subsystems or four subsystems, respectively.

Linearized model with two subsystems. In the following, the overall system is subdivided into two subsystems, representing the dynamics of the level and the temperature in the reactors TB and TS, respectively. Hence, the subsystem states are given by

$$\boldsymbol{x}_1 = \begin{pmatrix} l_{\mathrm{B}} \\ \vartheta_{\mathrm{B}} \end{pmatrix}, \quad \boldsymbol{x}_2 = \begin{pmatrix} l_{\mathrm{S}} \\ \vartheta_{\mathrm{S}} \end{pmatrix}.$$

The nonlinear model (A.1), (A.2f) is linearized around the operating point

$$\bar{\boldsymbol{x}}_1 = \begin{pmatrix} \bar{l}_{\mathrm{B}} \\ \bar{\vartheta}_{\mathrm{B}} \end{pmatrix} = \begin{pmatrix} 0.33\,\mathrm{m} \\ 294.7\,\mathrm{K} \end{pmatrix}, \quad \bar{\boldsymbol{x}}_2 = \begin{pmatrix} \bar{l}_{\mathrm{S}} \\ \bar{\vartheta}_{\mathrm{S}} \end{pmatrix} = \begin{pmatrix} 0.34\,\mathrm{m} \\ 300.2\,\mathrm{K} \end{pmatrix} \tag{A.5}$$

with

$$\bar{\boldsymbol{u}}_1 = \begin{pmatrix} \bar{u}_{\mathrm{T1}} \\ \bar{u}_{\mathrm{CU}} \end{pmatrix} = \begin{pmatrix} 0.5 \\ 0.5 \end{pmatrix}, \quad \bar{\boldsymbol{u}}_2 = \begin{pmatrix} \bar{u}_{\mathrm{T3}} \\ \bar{u}_{\mathrm{H}} \end{pmatrix} = \begin{pmatrix} 0.5 \\ 0.5 \end{pmatrix}.$$

and the valve angles

$$u_{\mathrm{BS}} = 0.19, \quad u_{\mathrm{SB}} = 0.22, \quad u_{\mathrm{BW}} = 0.21, \quad u_{\mathrm{SW}} = 0.29 \tag{A.6}$$

which define the coupling strength between the processes in the reactors TB and TS. These settings yield the linearized model of the form (2.3), (2.4) for $i = 1, 2$ with

$$\boldsymbol{A}_1 = 10^{-3} \begin{pmatrix} -5.74 & 0 \\ -34.5 & -8.58 \end{pmatrix}, \quad \boldsymbol{A}_2 = 10^{-3} \begin{pmatrix} -5.00 & 0 \\ 39.2 & -5.58 \end{pmatrix} \tag{A.7a}$$

$$\boldsymbol{B}_1 = 10^{-3} \begin{pmatrix} 2.30 & 0 \\ 0 & -38.9 \end{pmatrix}, \quad \boldsymbol{B}_2 = 10^{-3} \begin{pmatrix} 2.59 & 0 \\ 0 & 35.0 \end{pmatrix} \tag{A.7b}$$

$$\boldsymbol{E}_1 = 10^{-3} \begin{pmatrix} 0 \\ 169 \end{pmatrix}, \quad \boldsymbol{E}_2 = 10^{-3} \begin{pmatrix} 1.16 \\ -20.7 \end{pmatrix} \tag{A.7c}$$

$$\boldsymbol{E}_{\mathrm{s1}} = 10^{-3} \begin{pmatrix} 2.42 & 0 \\ 43.9 & 5.44 \end{pmatrix}, \quad \boldsymbol{E}_{\mathrm{s2}} = 10^{-3} \begin{pmatrix} 2.85 & 0 \\ -46.5 & 5.58 \end{pmatrix} \tag{A.7d}$$

$$\boldsymbol{C}_{\mathrm{z1}} = \begin{pmatrix} 1 & 0 \\ 0 & 1 \end{pmatrix}, \quad \boldsymbol{C}_{\mathrm{z2}} = \begin{pmatrix} 1 & 0 \\ 0 & 1 \end{pmatrix} \tag{A.7e}$$

and the interconnection (2.4) with

$$\boldsymbol{L}_{12} = \begin{pmatrix} 1 & 0 \\ 0 & 1 \end{pmatrix}, \quad \boldsymbol{L}_{21} = \begin{pmatrix} 1 & 0 \\ 0 & 1 \end{pmatrix}. \tag{A.8}$$

Linearized model with four subsystems. This paragraph presents the parameters of the linearized model that is obtained by a separation of the overall system into four scalar subsystems with the states

$$x_1(t) = l_{\mathrm{B}}(t), \quad x_2(t) = \vartheta_{\mathrm{B}}(t), \quad x_3(t) = l_{\mathrm{S}}(t), \quad x_4(t) = \vartheta_{\mathrm{S}}(t).$$

The overall system is composed of the subsystems (2.3) with the following parameters:

$$\text{Parameters of } \Sigma_1 : \begin{cases} \boldsymbol{A}_1 = -10^{-3} \cdot 5.74, \quad \boldsymbol{B}_1 = 10^{-3} \cdot 2.30, \quad \boldsymbol{E}_1 \ = 0, \\ \boldsymbol{E}_{\mathrm{s}1} = 10^{-3} \cdot 2.42, \quad \boldsymbol{C}_{\mathrm{z}1} = 1 \end{cases} \tag{A.9a}$$

$$\text{Parameters of } \Sigma_2 : \begin{cases} \boldsymbol{A}_2 = -10^{-3} \cdot 8.58, \quad \boldsymbol{B}_2 = -10^{-3} \cdot 38.9, \quad \boldsymbol{E}_2 = 0.169, \\ \boldsymbol{E}_{\mathrm{s}2} = 10^{-3} \begin{pmatrix} -34.5 & 43.9 & 5.44 \end{pmatrix}, \quad \boldsymbol{C}_{\mathrm{z}2} = 1 \end{cases} \tag{A.9b}$$

$$\text{Parameters of } \Sigma_3 : \begin{cases} \boldsymbol{A}_3 = -10^{-3} \cdot 5.00, \quad \boldsymbol{B}_3 = 10^{-3} \cdot 2.59, \\ \boldsymbol{E}_3 = 10^{-3} \cdot 1.16, \quad \boldsymbol{E}_{\mathrm{s}3} = 10^{-3} \cdot 2.85, \quad \boldsymbol{C}_{\mathrm{z}3} = 1 \end{cases} \tag{A.9c}$$

$$\text{Parameters of } \Sigma_4 : \begin{cases} \boldsymbol{A}_4 = -10^{-3} \cdot 5.58, \quad \boldsymbol{B}_4 = 10^{-3} \cdot 35.0, \quad \boldsymbol{E}_4 = -10^{-3} \cdot 20.7, \\ \boldsymbol{E}_{\mathrm{s}4} = 10^{-3} \begin{pmatrix} -46.5 & 5.58 & 39.2 \end{pmatrix}, \quad \boldsymbol{C}_{\mathrm{z}4} = 1. \end{cases} \tag{A.9d}$$

The interconnection between the subsystems is represented by Eq. (2.4) with

$$\begin{pmatrix} s_1(t) \\ \boldsymbol{s}_2(t) \\ s_3(t) \\ \boldsymbol{s}_4(t) \end{pmatrix} = \underbrace{\begin{pmatrix} 0 & 0 & 1 & 0 \\ \begin{pmatrix} 1 \\ 0 \\ 0 \end{pmatrix} & \begin{pmatrix} 0 \\ 0 \\ 0 \end{pmatrix} & \begin{pmatrix} 0 \\ 1 \\ 0 \end{pmatrix} & \begin{pmatrix} 0 \\ 0 \\ 1 \end{pmatrix} \\ 1 & 0 & 0 & 0 \\ \begin{pmatrix} 1 \\ 0 \\ 0 \end{pmatrix} & \begin{pmatrix} 0 \\ 1 \\ 0 \end{pmatrix} & \begin{pmatrix} 0 \\ 0 \\ 1 \end{pmatrix} & \begin{pmatrix} 0 \\ 0 \\ 0 \end{pmatrix} \end{pmatrix}}_{=\boldsymbol{L}} \begin{pmatrix} z_1(t) \\ z_2(t) \\ z_3(t) \\ z_4(t) \end{pmatrix}. \tag{A.10}$$

A.3 Approximate models

This section presents the approximate models for the thermofluid process with four subsystems (A.9), (A.10). These approximate models are applied for the design of a distributed state-feedback controller as proposed in Sec. 7.2,

The approximate models are designed under the assumption that subsystem Σ_i is controlled by a continuous state feedback

$$\boldsymbol{u}_i(t) = -\boldsymbol{K}_{\mathrm{d}i} \boldsymbol{x}_i(t),$$

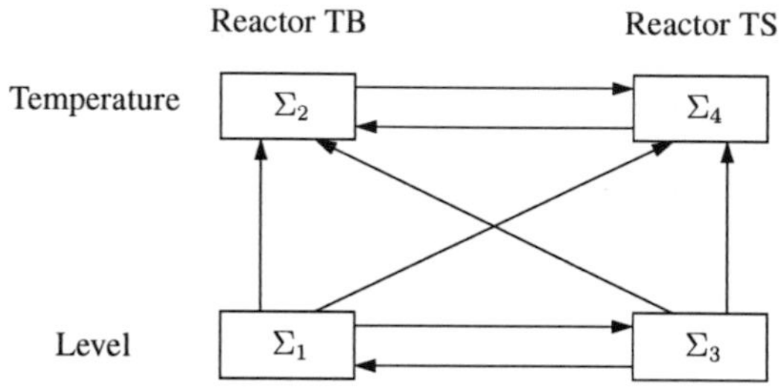

Figure 1.1: Interconnection of the thermofluid process with four subsystems

whereas in fact the control inputs are determined according to the distributed state feedback

$$\boldsymbol{u}_i(t) = -\boldsymbol{K}_{\mathrm{d}i}\boldsymbol{x}_i(t) - \boldsymbol{K}_{\mathrm{a}i}\boldsymbol{x}_{\mathrm{a}i}(t) = -\boldsymbol{K}_{\mathrm{d}i}\boldsymbol{x}_i(t) - \sum_{j\in\mathcal{P}_i}\boldsymbol{K}_{ij}\boldsymbol{x}_j(t).$$

The mismatch between the approximate model $\Sigma_{\mathrm{a}i}$ and the model that describes the behavior of the neighboring subsystems of Σ_i is aggregated in the residual model $\Sigma_{\mathrm{f}i}$.

Note that the sets $\mathcal{P}_i$ of predecessor subsystems implies the structure of the overall distributed

Table A.2: Parameters used for the design of the approximate models

Σ_i	$\mathcal{P}_i$	$\mathcal{S}_i$	$\boldsymbol{T}_{ij}$ as in (7.4)	$\boldsymbol{C}_{ji}$ as in (7.5)
Σ_1	$\{3\}$	$\{2,3,4\}$	$\boldsymbol{T}_{13} = 1$	$\boldsymbol{C}_{31} = 1$
Σ_2	$\{1,3,4\}$	$\{4\}$	$\boldsymbol{T}_{21} = \begin{pmatrix}1 & 0 & 0\end{pmatrix}^\top$	$\boldsymbol{C}_{12} = \begin{pmatrix}1 & 0 & 0\end{pmatrix}$
			$\boldsymbol{T}_{23} = \begin{pmatrix}0 & 1 & 0\end{pmatrix}^\top$	$\boldsymbol{C}_{32} = \begin{pmatrix}0 & 1 & 0\end{pmatrix}$
			$\boldsymbol{T}_{24} = \begin{pmatrix}0 & 0 & 1\end{pmatrix}^\top$	$\boldsymbol{C}_{42} = \begin{pmatrix}0 & 0 & 1\end{pmatrix}$
Σ_3	$\{1\}$	$\{1,2,4\}$	$\boldsymbol{T}_{31} = 1$	$\boldsymbol{C}_{13} = 1$
Σ_4	$\{1,2,3\}$	$\{2\}$	$\boldsymbol{T}_{41} = \begin{pmatrix}1 & 0 & 0\end{pmatrix}^\top$	$\boldsymbol{C}_{14} = \begin{pmatrix}1 & 0 & 0\end{pmatrix}$
			$\boldsymbol{T}_{42} = \begin{pmatrix}0 & 1 & 0\end{pmatrix}^\top$	$\boldsymbol{C}_{24} = \begin{pmatrix}0 & 1 & 0\end{pmatrix}$
			$\boldsymbol{T}_{43} = \begin{pmatrix}0 & 0 & 1\end{pmatrix}^\top$	$\boldsymbol{C}_{34} = \begin{pmatrix}0 & 0 & 1\end{pmatrix}$

state-feedback gain, which for this example is

$$K = \begin{pmatrix} K_{d1} & 0 & K_{13} & 0 \\ K_{21} & K_{d2} & K_{23} & K_{24} \\ K_{31} & 0 & K_{d3} & 0 \\ K_{41} & K_{42} & K_{43} & K_{d4} \end{pmatrix}.$$

Before the approximate models are presented, first consider Fig. 1.1 that illustrates the structure of the interconnection between the four subsystems of the thermofluid process. The set $\mathcal{P}_i$ of predecessor subsystems and the set $\mathcal{S}_i$ of successor subsystems for all subsystems are summarized in Table A.2. In addition to these sets the table also lists the choices for matrices $\boldsymbol{T}_{ij}$ and the corresponding $\boldsymbol{C}_{ji}$ applied for the transformations (7.4) and (7.5), respectively.

The approximate models $\Sigma_{\mathrm{a}i}$ (together with their corresponding residual model $\Sigma_{\mathrm{f}i}$) are presented in the following.

Approximate model Σ_{a1}. With the state

$$x_{\mathrm{a1}}(t) = x_3(t)$$

the approximate model Σ_{a1} is described by the state-space model

$$\Sigma_{\mathrm{a1}}: \begin{cases} \dot{x}_{\mathrm{a1}}(t) = (\boldsymbol{A}_3 - \boldsymbol{B}_3\boldsymbol{K}_{\mathrm{d3}})\, x_{\mathrm{a1}}(t) + \boldsymbol{E}_3 d_{\mathrm{a1}}(t) + \boldsymbol{E}_{\mathrm{s3}}\boldsymbol{L}_{31}\boldsymbol{C}_{\mathrm{z1}} z_1(t) + 1 \cdot f_1(t) \\ s_1(t) = x_{\mathrm{a1}}(t) \\ v_1(t) = z_1(t) \end{cases}$$

where $d_{\mathrm{a1}}(t) = d_{\mathrm{F}}(t)$ holds and the residual model Σ_{f1} is represented by

$$\Sigma_{\mathrm{f1}}: \quad f_1(t) = -\boldsymbol{B}_3\boldsymbol{K}_{\mathrm{a3}} \cdot v_1(t),$$

where $\boldsymbol{K}_{\mathrm{a3}} = \boldsymbol{K}_{31}$ holds.

Approximate model Σ_{a2}. According to the choice of the matrices $\boldsymbol{T}_{ij}$ the state of the approximate model Σ_{a2} is chosen to be

$$\boldsymbol{x}_{\mathrm{a2}}(t) = \begin{pmatrix} x_1(t) & x_2(t) & x_3(t) \end{pmatrix}^{\mathsf{T}}.$$

This transformation yields the approximate model

$$
\Sigma_{a2}: \begin{cases} \dot{\boldsymbol{x}}_{a2}(t) = \begin{pmatrix} \boldsymbol{A}_1 - \boldsymbol{B}_1\boldsymbol{K}_{d1} & \boldsymbol{E}_{s1}\boldsymbol{L}_{13}\boldsymbol{C}_{z3} & 0 \\ \boldsymbol{E}_{s3}\boldsymbol{L}_{31}\boldsymbol{C}_{z1} & \boldsymbol{A}_3 - \boldsymbol{B}_3\boldsymbol{K}_{d3} & 0 \\ \boldsymbol{E}_{s4}\boldsymbol{L}_{41}\boldsymbol{C}_{z1} & \boldsymbol{E}_{s4}\boldsymbol{L}_{43}\boldsymbol{C}_{z3} & \boldsymbol{A}_4 - \boldsymbol{B}_4\boldsymbol{K}_{d4} \end{pmatrix} \boldsymbol{x}_{a2}(t) \\ \qquad + \begin{pmatrix} \boldsymbol{E}_1 & 0 \\ 0 & \boldsymbol{E}_3 \\ 0 & \boldsymbol{E}_4 \end{pmatrix} \boldsymbol{d}_{a2}(t) + \begin{pmatrix} 0 \\ 0 \\ \boldsymbol{E}_{s4}\boldsymbol{L}_{42}\boldsymbol{C}_{z2} \end{pmatrix} z_2(t) + \begin{pmatrix} 1 & & \\ & 1 & \\ & & 1 \end{pmatrix} \boldsymbol{f}_2(t) \\ \boldsymbol{s}_2(t) = \boldsymbol{x}_{a2}(t) \\ \boldsymbol{v}_2(t) = \begin{pmatrix} \boldsymbol{I}_{(n_1+n_3+n_4)} \\ \boldsymbol{0} \end{pmatrix} \boldsymbol{x}_{a2}(t) + \begin{pmatrix} 0 \\ 1 \end{pmatrix} z_2(t) \end{cases}
$$

with the composed disturbance $\boldsymbol{d}_{a2}(t) = \begin{pmatrix} d_H(t) & d_F(t) \end{pmatrix}^\top$. The residual model Σ_{f2} is a static system that is described by the equation

$$
\Sigma_{f2}: \quad \boldsymbol{f}_2(t) = \begin{pmatrix} 0 & -\boldsymbol{B}_1\boldsymbol{K}_{13} & 0 & 0 \\ -\boldsymbol{B}_3\boldsymbol{K}_{31} & 0 & 0 & 0 \\ -\boldsymbol{B}_4\boldsymbol{K}_{41} & -\boldsymbol{B}_4\boldsymbol{K}_{43} & 0 & -\boldsymbol{B}_4\boldsymbol{K}_{42} \end{pmatrix} \boldsymbol{v}_2(t).
$$

Approximate model Σ_{a3}. With the state

$$
x_{a3}(t) = x_1(t)
$$

the approximate model Σ_{a3} is represented by the state-space model

$$
\Sigma_{a3}: \begin{cases} \dot{x}_{a3}(t) = (\boldsymbol{A}_1 - \boldsymbol{B}_1\boldsymbol{K}_{d1})\, x_{a3}(t) + \boldsymbol{E}_{s1}\boldsymbol{L}_{13}\boldsymbol{C}_{z3} z_3(t) + 1 \cdot f_3(t) \\ s_3(t) = x_{a3}(t) \\ v_3(t) = z_3(t). \end{cases}
$$

The residual model Σ_{f3} is given by

$$
\Sigma_{f3}: \quad f_3(t) = -\boldsymbol{B}_1\boldsymbol{K}_{a1} \cdot v_3(t)
$$

with $\boldsymbol{K}_{a1} = \boldsymbol{K}_{13}$.

Approximate model Σ_{a4}. The approximate model Σ_{a4} is obtained using the transformation

$$\boldsymbol{x}_{a4}(t) = \begin{pmatrix} x_1(t) & x_2(t) & x_3(t) \end{pmatrix}^\top$$

which yields

$$\Sigma_{a4}: \begin{cases} \dot{\boldsymbol{x}}_{a4}(t) = \begin{pmatrix} \boldsymbol{A}_1 - \boldsymbol{B}_1\boldsymbol{K}_{d1} & 0 & \boldsymbol{E}_{s1}\boldsymbol{L}_{13}\boldsymbol{C}_{z3} \\ \boldsymbol{E}_{s2}\boldsymbol{L}_{21}\boldsymbol{C}_{z1} & \boldsymbol{A}_2 - \boldsymbol{B}_2\boldsymbol{K}_{d2} & \boldsymbol{E}_{s2}\boldsymbol{L}_{23}\boldsymbol{C}_{z3} \\ \boldsymbol{E}_{s3}\boldsymbol{L}_{31}\boldsymbol{C}_{z1} & 0 & \boldsymbol{A}_3 - \boldsymbol{B}_3\boldsymbol{K}_{d3} \end{pmatrix} \boldsymbol{x}_{a4}(t) \\ \qquad + \begin{pmatrix} \boldsymbol{E}_1 & 0 \\ \boldsymbol{E}_2 & 0 \\ 0 & \boldsymbol{E}_3 \end{pmatrix} \boldsymbol{d}_{a4}(t) + \begin{pmatrix} 0 \\ \boldsymbol{E}_{s2}\boldsymbol{L}_{24}\boldsymbol{C}_{z4} \\ 0 \end{pmatrix} z_4(t) + \begin{pmatrix} 1 & & \\ & 1 & \\ & & 1 \end{pmatrix} \boldsymbol{f}_4(t) \\ \boldsymbol{s}_4(t) = \boldsymbol{x}_{a4}(t) \\ \boldsymbol{v}_4(t) = \begin{pmatrix} \boldsymbol{I}_{(n_1+n_2+n_3)} \\ \boldsymbol{0} \end{pmatrix} \boldsymbol{x}_{a4}(t) + \begin{pmatrix} 0 \\ 1 \end{pmatrix} z_4(t) \end{cases}$$

with the disturbance $\boldsymbol{d}_{a4}(t) = \begin{pmatrix} d_H(t) & d_F(t) \end{pmatrix}^\top$. This choice of the approximate model leads to the residual model

$$\Sigma_{f4}: \quad \boldsymbol{f}_4(t) = \begin{pmatrix} 0 & 0 & -\boldsymbol{B}_1\boldsymbol{K}_{13} & 0 \\ -\boldsymbol{B}_2\boldsymbol{K}_{21} & 0 & -\boldsymbol{B}_2\boldsymbol{K}_{23} & -\boldsymbol{B}_2\boldsymbol{K}_{24} \\ -\boldsymbol{B}_3\boldsymbol{K}_{31} & 0 & 0 & 0 \end{pmatrix} \boldsymbol{v}_4(t).$$

Note that in all cases the residual model Σ_{fi} is a static model, representing the error that is made by designing approximate models based on the assumption that the subsystems are controlled by a decentralized state feedback, whereas in fact all subsystems are controlled by a distributed state feedback.

B Proofs

B.1 Proof of proposition 6.1

Consider the difference state $\boldsymbol{x}_{\Delta i}(t)$ which is described by the state-space model (6.9). Let Eq. (6.17) hold and assume that, in contrast to the analysis in Sec. 6.2.5, the disturbance $\boldsymbol{d}_i(t)$ is not absent. Then (6.9) yields

$$\dot{\boldsymbol{x}}_{\Delta i}(t) = \boldsymbol{A}_i \boldsymbol{x}_{\Delta i}(t) + \boldsymbol{E}_i \boldsymbol{d}_i(t) + \boldsymbol{E}_{\mathrm{s}i}\big(\bar{\boldsymbol{s}}_i - \hat{\boldsymbol{s}}_i(t_{k_i})\big), \quad \boldsymbol{x}_{\Delta i}(t_{k_i}^+) = \boldsymbol{0}$$

which describes the difference state $\boldsymbol{x}_{\Delta i}(t)$ for the time $t \in [t_{k_i}, t_{k_i+1})$ where the constant estimate $\hat{\boldsymbol{s}}_i(t) = \hat{\boldsymbol{s}}_i(t_{k_i})$ is applied. From this model the equation

$$\begin{aligned}
\boldsymbol{x}_{\Delta i}(t) &= \int_{t_{k_i}}^{t} \mathrm{e}^{\boldsymbol{A}_i(t-\tau)} \boldsymbol{E}_i \boldsymbol{d}_i(\tau)\mathrm{d}\tau + \int_{t_{k_i}}^{t} \mathrm{e}^{\boldsymbol{A}_i(t-\tau)} \boldsymbol{E}_{\mathrm{s}i}\big(\bar{\boldsymbol{s}}_i - \hat{\boldsymbol{s}}_i(t_{k_i})\big)\mathrm{d}\tau \\
&= \int_{t_{k_i}}^{t} \mathrm{e}^{\boldsymbol{A}_i(t-\tau)} \boldsymbol{E}_i \boldsymbol{d}_i(\tau)\mathrm{d}\tau + \boldsymbol{A}_i^{-1}\left(\mathrm{e}^{\boldsymbol{A}_i(t-t_{k_i})} - \boldsymbol{I}_{n_i}\right)\boldsymbol{E}_{\mathrm{s}i}\big(\bar{\boldsymbol{s}}_i - \hat{\boldsymbol{s}}_i(t_{k_i})\big).
\end{aligned}$$

is obtained. At the next event time t_{k_i+1}

$$\begin{aligned}
\boldsymbol{x}_{\Delta i}(t_{k_i+1}) = \int_{t_{k_i}}^{t_{k_i+1}} & \mathrm{e}^{\boldsymbol{A}_i(t_{k_i+1}-\tau)} \boldsymbol{E}_i \boldsymbol{d}_i(\tau)\mathrm{d}\tau \\
& + \boldsymbol{A}_i^{-1}\left(\mathrm{e}^{\boldsymbol{A}_i(t_{k_i+1}-t_{k_i})} - \boldsymbol{I}_{n_i}\right)\boldsymbol{E}_{\mathrm{s}i}\big(\bar{\boldsymbol{s}}_i - \hat{\boldsymbol{s}}_i(t_{k_i})\big)
\end{aligned}$$

holds, which yields

$$\begin{aligned}
&\hat{\boldsymbol{s}}_i(t_{k_i}) + \left(\boldsymbol{A}_i^{-1}\left(\mathrm{e}^{\boldsymbol{A}_i(t_{k_i+1}-t_{k_i})} - \boldsymbol{I}_{n_i}\right)\boldsymbol{E}_{\mathrm{s}i}\right)^{+} \boldsymbol{x}_{\Delta i}(t_{k_i+1}) \\
&\quad = \bar{\boldsymbol{s}}_i + \left(\boldsymbol{A}_i^{-1}\left(\mathrm{e}^{\boldsymbol{A}_i(t_{k_i+1}-t_{k_i})} - \boldsymbol{I}_{n_i}\right)\boldsymbol{E}_{\mathrm{s}i}\right)^{+} \int_{t_{k_i}}^{t_{k_i+1}} \mathrm{e}^{\boldsymbol{A}_i(t_{k_i+1}-\tau)} \boldsymbol{E}_i \boldsymbol{d}_i(\tau)\mathrm{d}\tau.
\end{aligned}$$

By comparing the last equation with the estimation (6.18) it can be seen that the actual estimate $\hat{\boldsymbol{s}}_i(t_{k_i+1})$ is given by

$$\hat{\boldsymbol{s}}_i(t_{k_i+1}) := \bar{\boldsymbol{s}}_i(t_{k_i}) + \boldsymbol{\varepsilon}_i(t_{k_i+1})$$

where

$$\boldsymbol{\varepsilon}_i(t_{k_i+1}) = \left(\boldsymbol{A}_i^{-1}\left(\mathrm{e}^{\boldsymbol{A}_i(t_{k_i+1}-t_{k_i})} - \boldsymbol{I}_{n_i}\right)\boldsymbol{E}_{\mathrm{s}i}\right)^+ \int_{t_{k_i}}^{t_{k_i+1}} \mathrm{e}^{\boldsymbol{A}_i(t_{k_i+1}-\tau)}\boldsymbol{E}_i\boldsymbol{d}_i(\tau)\mathrm{d}\tau.$$

represents the estimation error, which completes the proof. □

B.2 Proof of proposition 6.2

The following analysis derives the estimation error (6.27) that emerges in the dynamic approach to the coupling estimation if assumption (6.16) is not fulfilled and subsystem Σ_i is disturbed. In that case the model (6.21) must be extended by the disturbance input which yields

$$\begin{aligned} \dot{\boldsymbol{x}}_i(t) &= \boldsymbol{A}_i\boldsymbol{x}_i(t) - \boldsymbol{B}_i\boldsymbol{K}_{\mathrm{d}i}\boldsymbol{x}_{\mathrm{s}i}(t) + \boldsymbol{E}_i\boldsymbol{d}_i(t) + \boldsymbol{E}_{\mathrm{s}i}\boldsymbol{s}_i(t), \qquad && \boldsymbol{x}_i(0) = \boldsymbol{x}_{0i} \\ \dot{\boldsymbol{s}}_i(t) &= \boldsymbol{A}_{\mathrm{s}i}\boldsymbol{s}_i(t), && \boldsymbol{s}_i(0) = \boldsymbol{s}_{i,0}. \end{aligned}$$

Then, for the difference system the model

$$\begin{aligned} \frac{\mathrm{d}}{\mathrm{d}t}\begin{pmatrix}\boldsymbol{x}_{\Delta i}(t)\\ \boldsymbol{s}_{\Delta i}(t)\end{pmatrix} &= \begin{pmatrix}\boldsymbol{A}_i & \boldsymbol{E}_{\mathrm{s}i}\\ \boldsymbol{O} & \boldsymbol{A}_{\mathrm{s}i}\end{pmatrix}\begin{pmatrix}\boldsymbol{x}_{\Delta i}(t)\\ \boldsymbol{s}_{\Delta i}(t)\end{pmatrix} + \begin{pmatrix}\boldsymbol{E}_i\\ \boldsymbol{O}\end{pmatrix}\boldsymbol{d}_i(t), \\ \begin{pmatrix}\boldsymbol{x}_{\Delta i}(t_{k_i}^+)\\ \boldsymbol{s}_{\Delta i}(t_{k_i}^+)\end{pmatrix} &= \begin{pmatrix}0\\ \boldsymbol{s}_i(t_{k_i}) - \hat{\boldsymbol{s}}_{i,k}\end{pmatrix} \end{aligned}$$

is obtained from which

$$\begin{aligned} \boldsymbol{x}_{\Delta i}(t) = {} & \begin{pmatrix}\boldsymbol{I}_{n_i} & \boldsymbol{O}\end{pmatrix}\mathrm{e}^{\boldsymbol{R}_i(t-t_{k_i})}\begin{pmatrix}\boldsymbol{O}\\ \boldsymbol{I}_{q_i}\end{pmatrix}\left(\boldsymbol{s}_i(t_{k_i}) - \hat{\boldsymbol{s}}_{i,k}\right) \\ & + \int_{t_{k_i}}^{t}\begin{pmatrix}\boldsymbol{I}_{n_i} & \boldsymbol{O}\end{pmatrix}\mathrm{e}^{\boldsymbol{R}_i(t-\tau)}\begin{pmatrix}\boldsymbol{E}_i\\ \boldsymbol{O}\end{pmatrix}\boldsymbol{d}_i(\tau)\mathrm{d}\tau \end{aligned}$$

follows with the matrix $\boldsymbol{R}_i$ given in (6.24). Solving the last equation at the next event time $t = t_{k_i+1}$ for the expression on the right-hand side of Eq. (6.25) yields

$$\begin{aligned}\hat{\boldsymbol{s}}_{i,k} &+ \left(\bar{\boldsymbol{R}}_i(t_{k_i+1} - t_{k_i})\right)^+ \boldsymbol{x}_{\Delta i}(t_{k_i+1}) \\ &= \boldsymbol{s}_i(t_{k_i}) + \left(\bar{\boldsymbol{R}}_i(t_{k_i+1} - t_{k_i})\right)^+ \int_{t_{k_i}}^{t_{k_i+1}} \begin{pmatrix} \boldsymbol{I}_{n_i} & \boldsymbol{O} \end{pmatrix} \mathrm{e}^{\boldsymbol{R}_i(t-\tau)} \begin{pmatrix} \boldsymbol{E}_i \\ \boldsymbol{O} \end{pmatrix} \boldsymbol{d}_i(\tau)\mathrm{d}\tau.\end{aligned}$$

Thus, Eq. (6.25) determines the initial value $\boldsymbol{s}_i(t_{k_i})$ plus an error term that depends upon the disturbance $\boldsymbol{d}_i(t)$ in the interval $[t_{k_i}, t_{k_i+1})$. Now observe that the determination of the new initial condition according to Eq. (6.26) results in

$$\begin{aligned}\hat{\boldsymbol{s}}_{i,k_i+1} &:= \mathrm{e}^{\boldsymbol{A}_{\mathrm{s}i}(t_{k_i+1} - t_{k_i})} \left(\hat{\boldsymbol{s}}_{i,k} + \left(\bar{\boldsymbol{R}}_i(t_{k_i+1} - t_{k_i})\right)^+ \boldsymbol{x}_{\Delta i}(t_{k_i+1})\right) \\ &= \mathrm{e}^{\boldsymbol{A}_{\mathrm{s}i}(t_{k_i+1} - t_{k_i})} \boldsymbol{s}_i(t_{k_i}) + \boldsymbol{\varepsilon}(t_{k_i+1})\end{aligned}$$

with the estimation error $\boldsymbol{\varepsilon}(t_{k_i+1})$ given in Eq. (6.27). □

B.3 Proof of Theorem 6.3

Consider the controlled reference systems (6.53) which yields

$$\boldsymbol{x}_{\mathrm{r}i}(t) = \boldsymbol{F}_i(t)\boldsymbol{x}_{0i} + \boldsymbol{G}_{\mathrm{xd}i}(t) * \boldsymbol{d}_i + \boldsymbol{G}_{\mathrm{xs}i} * \boldsymbol{s}_{\mathrm{r}i}$$

with the matrices $\boldsymbol{F}_i(t)$, $\boldsymbol{G}_{\mathrm{xd}i}(t)$ and $\boldsymbol{G}_{\mathrm{xs}i}(t)$ given in (6.42). Hence, the system

$$\boldsymbol{r}_{\mathrm{r}i}(t) = \bar{\boldsymbol{F}}_i(t)\,|\boldsymbol{x}_{0i}| + \bar{\boldsymbol{G}}_{\mathrm{xd}i} * |\boldsymbol{d}_i| + \bar{\boldsymbol{G}}_{\mathrm{xs}i} * |\boldsymbol{s}_{\mathrm{r}i}| \tag{B.1}$$

with $\bar{\boldsymbol{F}}_i(t) = |\boldsymbol{F}_i(t)|$, $\bar{\boldsymbol{G}}_{\mathrm{xd}i}(t) = |\boldsymbol{G}_{\mathrm{xd}i}(t)|$ and $\bar{\boldsymbol{G}}_{\mathrm{xs}i}(t) = |\boldsymbol{G}_{\mathrm{xs}i}(t)|$ is a comparison system for the original system (6.53).

In order to investigate asymptotic stability of the overall control system, set

$$\begin{aligned}\boldsymbol{r}_{\mathrm{r}}(t) &:= \begin{pmatrix} \boldsymbol{r}_{\mathrm{r}1}^{\top}(t) & \dots & \boldsymbol{r}_{\mathrm{r}N}^{\top}(t) \end{pmatrix}^{\top}, \\ \boldsymbol{s}_{\mathrm{r}}(t) &:= \begin{pmatrix} \boldsymbol{s}_{\mathrm{r}1}^{\top}(t) & \dots & \boldsymbol{s}_{\mathrm{r}N}^{\top}(t) \end{pmatrix}^{\top}\end{aligned}$$

and

$$\bar{\boldsymbol{F}}(t) := \operatorname{diag}\left(\bar{\boldsymbol{F}}_1(t), \ldots, \bar{\boldsymbol{F}}_N(t)\right),$$
$$\bar{\boldsymbol{G}}_{\mathrm{xd}}(t) := \operatorname{diag}\left(\bar{\boldsymbol{G}}_{\mathrm{xd}1}(t), \ldots, \bar{\boldsymbol{G}}_{\mathrm{xd}N}(t)\right),$$
$$\bar{\boldsymbol{G}}_{\mathrm{xs}}(t) := \operatorname{diag}\left(\bar{\boldsymbol{G}}_{\mathrm{xs}1}(t), \ldots, \bar{\boldsymbol{G}}_{\mathrm{xs}N}(t)\right).$$

Then the interconnection of the comparison systems (B.1) according to

$$|\boldsymbol{s}_{\mathrm{r}}(t)| \leq |\boldsymbol{L}|\,|\boldsymbol{z}_{\mathrm{r}}(t)| \leq |\boldsymbol{L}|\,|\boldsymbol{C}_{\mathrm{z}}|\,|\boldsymbol{x}_{\mathrm{r}}(t)|$$

yields the comparison system for the interconnected control loops

$$\begin{aligned}\boldsymbol{r}_{\mathrm{r}}(t) &= \bar{\boldsymbol{F}}(t)\,|\boldsymbol{x}_0| + \bar{\boldsymbol{G}}_{\mathrm{xd}} * |\boldsymbol{d}| + \bar{\boldsymbol{G}}_{\mathrm{xs}} * |\boldsymbol{s}_{\mathrm{r}}| \\ &= \bar{\boldsymbol{F}}(t)\,|\boldsymbol{x}_0| + \bar{\boldsymbol{G}}_{\mathrm{xd}} * |\boldsymbol{d}| + \bar{\boldsymbol{G}}_{\mathrm{xs}}\,|\boldsymbol{L}|\,|\boldsymbol{C}_{\mathrm{z}}| * |\boldsymbol{x}_{\mathrm{r}}| \geq |\boldsymbol{x}_{\mathrm{r}}(t)| \end{aligned} \tag{B.2}$$

with $\boldsymbol{C}_{\mathrm{z}} = \operatorname{diag}\left(\boldsymbol{C}_{\mathrm{z}1}, \ldots, \boldsymbol{C}_{\mathrm{z}N}\right)$. Inequality (B.2) is an implicit bound on the overall state $\boldsymbol{x}_{\mathrm{r}}(t)$. From the comparison principle, [113], it is known that if the condition

$$\lambda_{\mathrm{p}}\left(\int_0^{\infty} \bar{\boldsymbol{G}}_{\mathrm{xs}}(t)\,|\boldsymbol{L}|\,|\boldsymbol{C}_{\mathrm{z}}|\,\mathrm{d}t\right) < 1 \tag{B.3}$$

is satisfied, the impulse response matrix

$$\boldsymbol{G}(t) = \delta(t)\boldsymbol{I}_n + \bar{\boldsymbol{G}}_{\mathrm{xs}}\,|\boldsymbol{L}|\,|\boldsymbol{C}_{\mathrm{z}}| * \boldsymbol{G}$$

exists and is non-negative and, hence, (B.2) can be rewritten in explicit form as

$$\boldsymbol{r}_{\mathrm{r}}(t) = \boldsymbol{G} * \left(\bar{\boldsymbol{F}}\,|\boldsymbol{x}_0| + \bar{\boldsymbol{G}}_{\mathrm{xd}} * |\boldsymbol{d}|\right) \geq |\boldsymbol{x}_{\mathrm{r}}(t)|\,. \tag{B.4}$$

The system (B.4) is a comparison system for the interconnected continuous control loops (6.53), (6.54). By virtue of the condition (B.3), $\boldsymbol{r}_{\mathrm{r}}(t)$ is known to be bounded which implies the stability of the interconnected continuous control loops (6.53), (6.54). □

B.4 Proof of Theorem 6.5

Consider the comparison system

$$\boldsymbol{r}_{\mathrm{r}}(t) = \boldsymbol{G} * \left(\bar{\boldsymbol{F}}\,|\boldsymbol{x}_0| + \bar{\boldsymbol{G}}_{\mathrm{xd}} * |\boldsymbol{d}|\right) \geq |\boldsymbol{x}_{\mathrm{r}}(t)|$$

for the interconnected reference systems (6.53), (6.54). With the bound

$$\bar{\boldsymbol{G}}_{\mathrm{xd}} * |\boldsymbol{d}| \leq \int_0^\infty \bar{\boldsymbol{G}}_{\mathrm{xd}}(t)\mathrm{d}t \cdot \bar{\boldsymbol{d}} =: \boldsymbol{M}_{\mathrm{xd}} \cdot \bar{\boldsymbol{d}}$$

the last equation yields

$$|\boldsymbol{x}_{\mathrm{r}}(t)| \leq \boldsymbol{G} * \bar{\boldsymbol{F}}(t)\,|\boldsymbol{x}_0| + \boldsymbol{G} * \boldsymbol{M}_{\mathrm{xd}}\bar{\boldsymbol{d}} = \boldsymbol{r}_{\mathrm{r}}(t). \tag{B.5}$$

Now, observe that the term which depends upon the initial state $\boldsymbol{x}_0$ vanishes for $t \to \infty$. Hence, from the previous reference system the bound

$$\limsup_{t\to\infty} |\boldsymbol{x}_{\mathrm{r}}(t)| \leq \lim_{t\to\infty} \boldsymbol{r}_{\mathrm{r}}(t) = \int_0^\infty \boldsymbol{G}(t)\mathrm{d}t\ \boldsymbol{M}_{\mathrm{xd}}\bar{\boldsymbol{d}} =: \boldsymbol{b}_{\mathrm{r}}$$

is obtained. Using Eq. (6.52)

$$\boldsymbol{b}_{\mathrm{r}} = \hat{\boldsymbol{G}}^{-1}\boldsymbol{M}_{\mathrm{xd}}\bar{\boldsymbol{d}}$$

follows. Hence, for $t \to \infty$ the signal $\boldsymbol{r}_{\mathrm{r}}(t)$ of the comparison system (B.5) converges to the set

$$\mathcal{A}_{\mathrm{r}} := \left\{\boldsymbol{x}_{\mathrm{r}} \in \mathbb{R}^n \;\middle|\; |\boldsymbol{x}_{\mathrm{r}}| \leq \boldsymbol{b}_{\mathrm{r}}\right\}.$$

Since $\boldsymbol{r}_{\mathrm{r}}(t)$ is known to be a bound on the state $\boldsymbol{x}_{\mathrm{r}}(t)$ of the interconnected reference systems (6.53), (6.54), it can be concluded that the state $\boldsymbol{x}_{\mathrm{r}}(t)$ converges to the set $\mathcal{A}_{\mathrm{r}}$ as well. □

B.5 Proof of Lemma 6.2

Consider the interconnected event-based state-feedback loops (2.3), (2.4), (6.32), (6.34). For the time $t \in [t_{k_i}, t_{k_i+1})$ the difference state $\boldsymbol{x}_{\Delta i}(t) = \boldsymbol{x}_i(t) - \boldsymbol{x}_{\mathrm{s}i}(t)$ is described by the state-space model

$$\dot{\boldsymbol{x}}_{\Delta i}(t) = \boldsymbol{A}_i\boldsymbol{x}_{\Delta i}(t) + \boldsymbol{E}_i\left(\boldsymbol{d}_i(t) - \hat{\boldsymbol{d}}_{i,k_i}\right) + \boldsymbol{E}_{\mathrm{s}i}\boldsymbol{s}_i(t), \quad \boldsymbol{x}_{\Delta i}(t_{k_i}^+) = \boldsymbol{0},$$

which yields

$$\boldsymbol{x}_{\Delta i}(t) = \int_{t_{k_i}}^t \mathrm{e}^{\boldsymbol{A}_i(t-\tau)}\left(\boldsymbol{E}_i\left(\boldsymbol{d}_i(\tau) - \hat{\boldsymbol{d}}_{i,k_i}\right) + \boldsymbol{E}_{\mathrm{s}i}\boldsymbol{s}_i(\tau)\right)\mathrm{d}\tau.$$

The minimum inter-event time $T_{\min i}$ for two consecutive events triggered by E_i is given by the minimum time t for which at least one row of the equation

$$|\boldsymbol{x}_{\Delta i}(t)| = \left| \int_0^t \mathrm{e}^{\boldsymbol{A}_i(t-\tau)} \left(\boldsymbol{E}_i \left(\boldsymbol{d}_i(\tau) - \hat{\boldsymbol{d}}_{i,k_i} \right) + \boldsymbol{E}_{\mathrm{s}i} \boldsymbol{s}_i(\tau) \right) \mathrm{d}\tau \right| = \bar{\boldsymbol{e}}_i$$

holds. The following investigation derives a bound $\bar{T}_i$ on the minimum inter-event time $T_{\min i}$. To this end assume that maximum error of the disturbance estimation satisfies the relation

$$\left| \boldsymbol{d}_i(t) - \hat{\boldsymbol{d}}_{i,k_i} \right| \leq \bar{\boldsymbol{d}}_{\Delta i}$$

for all $t \geq 0$ and all $k_i \in \mathbb{N}_0$ and that the coupling input $\boldsymbol{s}_i(t)$ is bounded for all $t \geq 0$. Then the difference state $\boldsymbol{x}_{\Delta i}(t)$ can be bounded as follows:

$$|\boldsymbol{x}_{\Delta i}(t)| \leq \int_0^t \left| \mathrm{e}^{\boldsymbol{A}_i \tau} \right| \mathrm{d}\tau \left(|\boldsymbol{E}_i| \, \bar{\boldsymbol{d}}_{\Delta i} + |\boldsymbol{E}_{\mathrm{s}i}| \cdot \sup_{t \geq 0} |\boldsymbol{s}_i(t)| \right).$$

In view of this estimation, a bound $\bar{T}_i$ on the minimum inter-event time $T_{\min i}$ is given by

$$\bar{T}_i := \arg\min_t \left\{ \int_0^t \left| \mathrm{e}^{\boldsymbol{A}_i t} \right| \mathrm{d}t \left(|\boldsymbol{E}_i| \, \bar{\boldsymbol{d}}_{\Delta i} + |\boldsymbol{E}_{\mathrm{s}i}| \cdot \sup_{t \geq 0} |\boldsymbol{s}_i(t)| \right) = \bar{\boldsymbol{e}}_i \right\},$$

which completes the proof. □

B.6 Proof of Theorem 7.1

The proposed analysis method makes use of comparison systems, which are explained in Sec. 6.3.3. The following equations

$$\boldsymbol{r}_{\mathrm{x}i}(t) = \boldsymbol{V}_{\mathrm{x0}i}(t) \, |\boldsymbol{x}_{\mathrm{e0}i}| + \boldsymbol{V}_{\mathrm{xd}i} * |\boldsymbol{d}_{\mathrm{e}i}| + \boldsymbol{V}_{\mathrm{xf}i} * |\boldsymbol{f}_i| \geq |\boldsymbol{x}_{\mathrm{e}i}(t)| \qquad \text{(B.6a)}$$

$$\boldsymbol{r}_{\mathrm{v}i}(t) = \boldsymbol{V}_{\mathrm{v0}i}(t) \, |\boldsymbol{x}_{\mathrm{e0}i}| + \boldsymbol{V}_{\mathrm{vd}i} * |\boldsymbol{d}_{\mathrm{e}i}| + \boldsymbol{V}_{\mathrm{vf}i} * |\boldsymbol{f}_i| \geq |\boldsymbol{v}_i(t)|, \qquad \text{(B.6b)}$$

with

$$\boldsymbol{V}_{\mathrm{x0}i}(t) = \left| \mathrm{e}^{\bar{\boldsymbol{A}}_{\mathrm{e}i} t} \right|, \qquad \boldsymbol{V}_{\mathrm{xd}i}(t) = \left| \mathrm{e}^{\bar{\boldsymbol{A}}_{\mathrm{e}i} t} \boldsymbol{E}_{\mathrm{e}i} \right|, \qquad \boldsymbol{V}_{\mathrm{xf}i}(t) = \left| \mathrm{e}^{\bar{\boldsymbol{A}}_{\mathrm{e}i} t} \boldsymbol{F}_{\mathrm{e}i} \right|,$$

$$\boldsymbol{V}_{\mathrm{v0}i}(t) = \left| \boldsymbol{H}_{\mathrm{e}i} \mathrm{e}^{\bar{\boldsymbol{A}}_{\mathrm{e}i} t} \right|, \quad \boldsymbol{V}_{\mathrm{vd}i}(t) = \left| \boldsymbol{H}_{\mathrm{e}i} \mathrm{e}^{\bar{\boldsymbol{A}}_{\mathrm{e}i} t} \boldsymbol{E}_{\mathrm{e}i} \right|, \quad \boldsymbol{V}_{\mathrm{vf}i}(t) = \left| \boldsymbol{H}_{\mathrm{e}i} \mathrm{e}^{\bar{\boldsymbol{A}}_{\mathrm{e}i} t} \boldsymbol{F}_{\mathrm{e}i} \right|$$

represent a comparison system for the extended subsystem (7.9) with the distributed state feedback (7.16). In the following analysis the assumption $\boldsymbol{x}_{\mathrm{e0}i} = \boldsymbol{0}$ is made, which is no loss of generality for the presented stability criterion but simplifies the derivation. Consider the interconnection of the comparison system (B.6) and the residual system (7.6), (7.8), as illustrated in Fig. 2.1. The signal $\boldsymbol{r}_{\mathrm{f}i}(t) \geq |\boldsymbol{f}_i(t)|$ for all $t \geq 0$ is represented by

$$\boldsymbol{r}_{\mathrm{f}i}(t) = \bar{\boldsymbol{G}}_{\mathrm{fv}i} * \boldsymbol{V}_{\mathrm{vd}i} * |\boldsymbol{d}_{\mathrm{e}i}| + \bar{\boldsymbol{G}}_{\mathrm{fd}i} * |\boldsymbol{d}_{\mathrm{f}i}| + \bar{\boldsymbol{G}}_{\mathrm{fv}i} * \boldsymbol{V}_{\mathrm{vf}i} * |\boldsymbol{f}_i| \geq |\boldsymbol{f}_i(t)| \tag{B.7}$$

which is an implicit bound on the signal $|\boldsymbol{f}_i(t)|$.

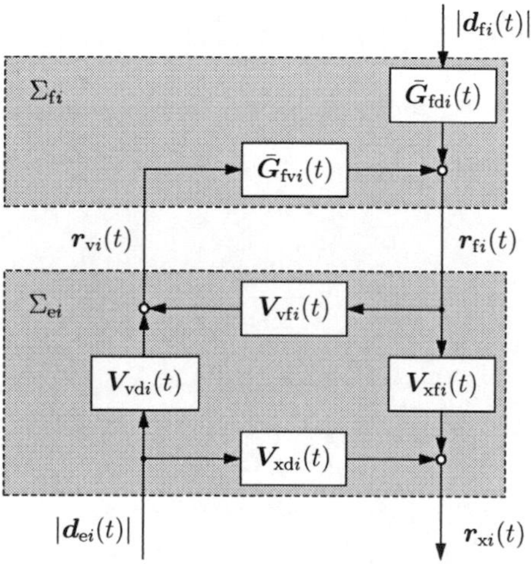

Figure 2.1: Interconnection of the comparison system (7.17) and the residual system (7.6), (7.8)

The structure of Eq. (B.7) is illustrated in Fig. 2.2, where the dependence of $\boldsymbol{r}_{\mathrm{f}i}(t)$ upon itself is represented by the feedback loop. An explicit bound on $|\boldsymbol{f}_i(t)|$ in terms of the signal $\boldsymbol{r}_{\mathrm{v}i}(t)$ can be obtained by means of the comparison principle, [113], where the basic idea is to find an impulse response matrix $\boldsymbol{G}_i(t)$ that describes the input/output behavior of the gray highlighted block in Fig. 2.2:

$$\boldsymbol{r}_{\mathrm{f}i}(t) = \boldsymbol{G}_i * \left(\bar{\boldsymbol{G}}_{\mathrm{fv}i} * \boldsymbol{V}_{\mathrm{vd}i} * |\boldsymbol{d}_{\mathrm{e}i}| + \bar{\boldsymbol{G}}_{\mathrm{fd}i} * |\boldsymbol{d}_{\mathrm{f}i}|\right) \geq |\boldsymbol{f}_i(t)|, \quad \forall\, t \geq 0. \tag{B.8}$$

The comparison principle says that the impulse response matrix

$$\boldsymbol{G}_i(t) = \delta(t)\boldsymbol{I} + \bar{\boldsymbol{G}}_{\mathrm{fv}i} * \boldsymbol{V}_{\mathrm{vf}i} * \boldsymbol{G}_i \tag{B.9}$$

exists, that means the inequality

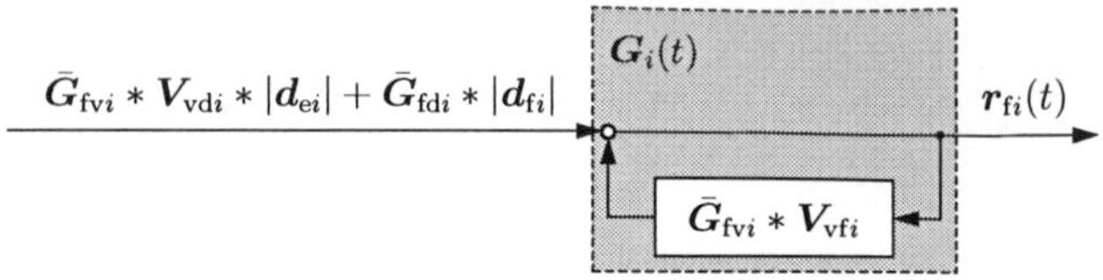

Figure 2.2: Block diagram representing Eq. (B.7)

$$\int_0^\infty \boldsymbol{G}_i(t)\mathrm{d}t = \boldsymbol{I} + \int_0^\infty \bar{\boldsymbol{G}}_{\text{fv}i} * \boldsymbol{V}_{\text{vf}i} * \boldsymbol{G}_i \mathrm{d}t$$
$$= \boldsymbol{I} + \int_0^\infty \bar{\boldsymbol{G}}_{\text{fv}i}(t)\mathrm{d}t \int_0^\infty \boldsymbol{V}_{\text{vf}i}(t)\mathrm{d}t \int_0^\infty \boldsymbol{G}_i(t)\mathrm{d}t < \infty$$

holds true if the condition

$$\lambda_{\text{P}}\left(\int_0^\infty \bar{\boldsymbol{G}}_{\text{fv}i}(t)\mathrm{d}t \int_0^\infty \boldsymbol{V}_{\text{vf}i}(t)\mathrm{d}t\right) < 1 \qquad \text{(B.10)}$$

is satisfied. On condition that relation (B.10) is fulfilled, the signal $\boldsymbol{r}_{\text{f}i}(t)$ remains bounded for all $t \geq 0$ which implies that $\boldsymbol{f}_i(t)$ is bounded, as well. Given that the disturbance inputs $\boldsymbol{d}_{\text{f}i}(t)$ and $\boldsymbol{d}_{\text{e}i}(t)$ are bounded by (7.7) or (7.11), respectively, and that the matrix $\bar{\boldsymbol{A}}_{\text{e}i}$ in (7.18) is Hurwitz by definition, the output $\boldsymbol{r}_{\text{x}i}(t)$ of the comparison system (7.17) is bounded for all $t \geq 0$. Since $\boldsymbol{r}_{\text{x}i}(t)$ represents a bound on the subsystem state $\boldsymbol{x}_i(t)$ the stability of the overall comparison system implies the stability of the (7.9) with the distributed state feedback (7.16) together with the residual model (7.6). □

B.7 Proof of Theorem 7.2

To begin with the proof of Theorem 7.2, observe that the state $\boldsymbol{x}_{\text{e}i}(t)$ of each controlled extended subsystem (7.9), (7.16) is bounded if the disturbance $\boldsymbol{d}_{\text{e}i}(t)$ as well as the signal $\boldsymbol{f}_i(t)$ are bounded. While the former is bounded by definition (cf. Eq. (7.11)) the latter is known to be bounded if the condition (7.19) holds which is fulfilled by assumption.

The next analysis derives the asymptotically stable set $\mathcal{A}_{\text{r}i}$ for the subsystems Σ_i. Consider Eq. (B.6a) from which

$$|\boldsymbol{x}_i(t)| = \begin{pmatrix}\boldsymbol{I}_{n_i} & \boldsymbol{O}\end{pmatrix} |\boldsymbol{x}_{\text{e}i}(t)| \leq \begin{pmatrix}\boldsymbol{I}_{n_i} & \boldsymbol{O}\end{pmatrix} \left(\boldsymbol{V}_{\text{x}0i}(t)\,|\boldsymbol{x}_{\text{e}0i}| + \boldsymbol{V}_{\text{xd}i} * \bar{\boldsymbol{d}}_{\text{e}i} + \boldsymbol{V}_{\text{xf}i} * |\boldsymbol{f}_i|\right)$$

follows, where the disturbance $\boldsymbol{d}_{\text{e}i}$ has been replaced by the bound given in (7.11). Note that for $t \to \infty$ the term that depends upon the initial state $\boldsymbol{x}_{\text{e}0i}$ vanishes. Hence, from the previous relation the bound

$$\limsup_{t\to\infty} |\boldsymbol{x}_i(t)| \leq \begin{pmatrix} \boldsymbol{I}_{n_i} & \boldsymbol{O} \end{pmatrix} \left(\int_0^\infty \boldsymbol{V}_{\mathrm{xd}i}(t)\mathrm{d}t \cdot \bar{\boldsymbol{d}}_{\mathrm{e}i} + \int_0^\infty \boldsymbol{V}_{\mathrm{xf}i}(t)\mathrm{d}t \cdot \bar{\boldsymbol{f}}_i \right) =: \boldsymbol{b}_{\mathrm{r}i} \quad \text{(B.11)}$$

follows with

$$\limsup_{t\to\infty} |\boldsymbol{f}_i(t)| \leq \bar{\boldsymbol{f}}_i(t).$$

From Eq. B.11 it can be inferred that subsystem Σ_i is practically stable with respect to the set

$$\mathcal{A}_{\mathrm{r}i} := \left\{ \boldsymbol{x}_i \in \mathbb{R}^{n_i} \;\middle|\; |\boldsymbol{x}_i| \leq \boldsymbol{b}_{\mathrm{r}i} \right\}. \quad \text{(B.12)}$$

The aim of the following analysis is to specify the bound $\bar{\boldsymbol{f}}_i$. Therefore, consider Eqs. (7.6), (B.6b) which yield

$$\boldsymbol{r}_{\mathrm{f}i}(t) = \bar{\boldsymbol{G}}_{\mathrm{fv}i} * (\boldsymbol{V}_{\mathrm{v}0i}\,|\boldsymbol{x}_{\mathrm{e}0i}| + \boldsymbol{V}_{\mathrm{vd}i} * |\boldsymbol{d}_{\mathrm{e}i}| + \boldsymbol{V}_{\mathrm{vf}i} * |\boldsymbol{f}_i|) + \bar{\boldsymbol{G}}_{\mathrm{fd}i} * |\boldsymbol{d}_{\mathrm{f}i}| \geq |\boldsymbol{f}_i(t)|\,.$$

Using the comparison principle this bound can be restated in the explicit form

$$\boldsymbol{r}_{\mathrm{f}i}(t) = \boldsymbol{G}_i * \left(\bar{\boldsymbol{G}}_{\mathrm{fv}i} * (\boldsymbol{V}_{\mathrm{v}0i}\,|\boldsymbol{x}_{\mathrm{e}0i}| + \boldsymbol{V}_{\mathrm{vd}i} * |\boldsymbol{d}_{\mathrm{e}i}|) + \bar{\boldsymbol{G}}_{\mathrm{fd}i} * |\boldsymbol{d}_{\mathrm{f}i}| \right) \geq |\boldsymbol{f}_i(t)|$$

with the impulse response matrix $\boldsymbol{G}_i$ given in (B.9). From the last relation the bound

$$\limsup_{t\to\infty} \boldsymbol{r}_{\mathrm{f}i}(t) = \int_0^\infty \boldsymbol{G}_i(t)\mathrm{d}t \left(\int_0^\infty \bar{\boldsymbol{G}}_{\mathrm{fv}i}(t)\mathrm{d}t \int_0^\infty \boldsymbol{V}_{\mathrm{vd}i}(t)\mathrm{d}t \cdot \bar{\boldsymbol{d}}_{\mathrm{e}i} + \int_0^\infty \bar{\boldsymbol{G}}_{\mathrm{fd}i}(t)\mathrm{d}t \cdot \bar{\boldsymbol{d}}_{\mathrm{f}i} \right) =: \bar{\boldsymbol{f}}_i \quad \text{(B.13)}$$

follows.

In summary, for $t \to \infty$ the state $\boldsymbol{x}_i(t)$ asymptotically converges to the set $\mathcal{A}_{\mathrm{r}i}$ defined in (B.12) with the ultimate bound $\boldsymbol{b}_{\mathrm{r}i}$ given in (B.11). The bound $\boldsymbol{b}_{\mathrm{r}i}$ is a function of the disturbance bound $\bar{\boldsymbol{d}}_{\mathrm{e}i}$ according to (7.11) and of the bound $\bar{\boldsymbol{f}}_i$ given in (B.13). This competes the proof of Theorem 7.2. □

C List of symbols

General conventions

- **Scalars** are represented by lower-case italic letters (x, u, d).
- **Vectors** are represented by lower-case bold italic letters ($\boldsymbol{x}, \boldsymbol{u}, \boldsymbol{d}$).
- **Matrices** are represented by upper-case bold italic letters ($\boldsymbol{A}, \boldsymbol{B}, \boldsymbol{E}$).

Indices

$(.)_{\max}$	Maximum value of a scalar
$(.)_{\min}$	Minimum value of a scalar
$(.)^{-1}$	Inverse of a matrix
$(.)^{+}$	Pseudoinverse of a matrix
$(.)^{\top}$	Transpose of a vector or matrix
$\dot{(.)}$	Time derivative of a signal
$(.)_0$	Initial value of a signal at time $t = 0$
$(.)_{ij}$	Block-diagonal entry of the i-th row and j-th column of a matrix
$(.)_{\mathrm{r}}$	Signal of the continuous-time reference system
$(.)_{\Delta}$	Difference signal

Systems

Σ	Overall system
Σ_i	Subsystem i
C	Control input generator
C_i	Control input generator of subsystem Σ_i

E	Event generator
E_i	Event generator of subsystem Σ_i
F	Event-based controller
F_i	Event-based controller of subsystem Σ_i
$\Sigma_{\mathrm{a}i}$	Approximate system for subsystem Σ_i
$\Sigma_{\mathrm{e}i}$	Extended subsystem for subsystem Σ_i

Scalars

n	Dimension of the state vector
m	Dimension of the input vector
p	Dimension of the disturbance vector
q	Dimension of the coupling input vector
r	Dimension of the coupling output vector
b	Ultimate bound (Used in the definition of the set $\mathcal{A}$)
$\bar{e}$	Event threshold
t	Time
t_k	Global event time
k	Global event counter
t_{k_i}	Event time instant triggered by event generator E_i
k_i	Counter for events triggered by event generator E_i
$t_{r_i(j)}$	Event that induces an information request from controller F_j triggered by controller F_i (In Chapter 7)
$r_i(j)$	Counter for information requests from controller F_j triggered by controller F_i (In Chapter 7)

Vectors

$\boldsymbol{x}$	State vector
$\boldsymbol{u}$	Input vector
$\boldsymbol{d}$	Disturbance vector
$\boldsymbol{s}$	Coupling input vector

$\boldsymbol{z}$	Coupling output vector
$\boldsymbol{v}$	Residual input vector
$\boldsymbol{f}$	Residual output vector
$\boldsymbol{b}$	Vectorial ultimate bound (Used in the definition of the set $\mathcal{A}$)
$\hat{\boldsymbol{d}}$	Disturbance estimate
$\bar{\boldsymbol{x}}$	Setpoint
$\boldsymbol{x}_{\mathrm{s}}$	Prediction of the state $\boldsymbol{x}(t)$ determined by the event-based controller
$\tilde{\boldsymbol{x}}_j^i$	Prediction of the state $\boldsymbol{x}_j(t)$ determined by the controller F_i (In Chapter 7)
$\boldsymbol{x}_\Delta$	Difference state ($\boldsymbol{x}_\Delta = \boldsymbol{x} - \boldsymbol{x}_{\mathrm{s}}$)
$\boldsymbol{x}_{\Delta j}^i$	Difference state ($\boldsymbol{x}_{\Delta j}^i = \boldsymbol{x}_j - \tilde{\boldsymbol{x}}_j^i$) (In Chapter 7)
$\bar{\boldsymbol{e}}$	Event threshold vector
$\boldsymbol{x}_{\mathrm{r}}$	State of the continuous-time reference system
$\boldsymbol{e}$	Approximation error ($\boldsymbol{e} = \boldsymbol{x} - \boldsymbol{x}_{\mathrm{r}}$)
$\mathbf{0}$	Null vector of appropriate dimension
$\mathbf{1}$	Vector composed of ones of appropriate dimension

Matrices

$\boldsymbol{A}$	System matrix
$\boldsymbol{B}$	Input matrix
$\boldsymbol{E}$	Disturbance input matrix
$\boldsymbol{E}_{\mathrm{s}}$	Coupling input matrix
$\boldsymbol{C}_{\mathrm{z}}$	Coupling output matrix
$\boldsymbol{L}$	Interconnection matrix
$\boldsymbol{K}$	Controller matrix
$\bar{\boldsymbol{A}}$	System matrix of a closed-loop system ($\bar{\boldsymbol{A}} = \boldsymbol{A} - \boldsymbol{B}\boldsymbol{K}$)
$\boldsymbol{I}_n$	Identity matrix of size n
$\boldsymbol{O}$	Null matrix of appropriate size

Sets

$\mathbb{N}$, $\mathbb{N}_0$	Set of natural numbers and set of natural numbers including 0
$\mathbb{R}$, $\mathbb{R}_+$	Set of real numbers and set of positive real numbers
$\mathbb{R}^n$	Set of real vectors with dimension n
$\mathbb{R}^{n\times m}$	Set of real matrices with n rows and m columns
$\mathcal{D}$	Set of disturbances
$\mathcal{T}$	Target set
$\mathcal{X}$	Constrained state-space
$\mathcal{A}$	Asymptotically stable set of a control system
$\mathcal{N}$	Set of subsystems
$\mathcal{P}_i$	Set of predecessor subsystems of subsystem Σ_i
$\mathcal{S}_i$	Set of successor subsystems of subsystem Σ_i

Persönliche Daten:

Name	Christian Stöcker
Geburtsdatum	18.02.1983
Geburtsort	Salzkotten

Bildungs- und Berufsweg:

Seit 05 / 2009	Wissenschaftlicher Mitarbeiter am Lehrstuhl für Automatisierungstechnik und Prozessinformatik (Prof. Dr.-Ing. Jan Lunze), Ruhr-Universität Bochum.
04 / 2007 - 03 / 2009	Studium der Elektrotechnik an der Ruhr-Universität Bochum mit den Schwerpunkten Regelungstechnik und Digitale Signalverarbeitung, Titel der Diplomarbeit: *Analyse überabtastender komplex-modulierter Filterbänke mit kleiner Gruppenlaufzeit*.
09 / 2003 - 02 / 2007	Studium der Elektro- und Informationstechnik an der Technischen Fachhochschule Georg Agricola, Titel der Diplomarbeit: *Beitrag zur Entwicklung einer Regelung für einen Verteilnetztransformator*.
06 / 2003 - 08 / 2003	Praktikum bei der Resol - Elektronische Regelungen GmbH, Hattingen.
09 / 2002 - 06 / 2003	Zivildienst in der Klinik Blankenstein, Hattingen.
08 / 1993 - 07 / 2002	Graf-Engelbert-Schule, Bochum.